Speaking Green with a Southern Accent

Speaking Green with a Southern Accent

Environmental Management and Innovation in the South

Edited by
Gerald Andrews Emison
and John C. Morris

LEXINGTON BOOKS
A division of
ROWMAN & LITTLEFIELD PUBLISHERS, INC.
Lanham • Boulder • New York • Toronto • Plymouth, UK

Published by Lexington Books
A division of Rowman & Littlefield Publishers, Inc.
A wholly owned subsidiary of The Rowman & Littlefield Publishing Group, Inc.
4501 Forbes Boulevard, Suite 200, Lanham, Maryland 20706
www.lexingtonbooks.com

Estover Road, Plymouth PL6 7PY, United Kingdom

British Library Cataloguing in Publication Information Available

Library of Congress Cataloging-in-Publication Data
Speaking green with a Southern accent : environmental management and
innovation in the South / edited by Gerald Andrews Emison and John C. Morris.
    p. cm.
  Includes bibliographical references and index.
  ISBN 978-0-7391-4651-4 (cloth : alk. paper) — ISBN 978-0-7391-4653-8
(electronic)
  1. Environmental management—Southern States. 2. Environmental planning—
Southern States. 3. Environmental policy—Southern States. 4. Southern
States—Environmental conditions. I. Emison, Gerald A. II. Morris, John C. (John
Charles), 1959–
  GE3015.U65S687 2010
  363.70975—dc22   2010027727

Printed in the United States of America

To Donna, who has never wavered in her support or cheerfulness

—GAE

To the memory of my parents,
Charles Howard Morris and Mary Baker Morris

—JCM

Contents

Tables and Figure

Acknowledgments

The genesis of this book began some years ago as a panel organized by Doug Goodman of Mississippi State University at a meeting of the Southern Political Science Association in Atlanta, Georgia. The feedback we received from the audience and the discussant was encouraging, and we embarked on the process to assemble a volume on environmental policy and management in southern states. Encouraged by friends and colleagues, we worked to assemble a group of both established and emerging scholars whose work was pertinent to our vision for this book. We are grateful to our contributing authors, all of whom have cheerfully responded to our seemingly endless requests for revisions and updates. This work is very much a reflection of their individual and collective efforts.

We also wish to acknowledge the contributions and assistance of several other people whose help proved invaluable in the production of this manuscript. Katie Neill at Old Dominion University provided invaluable editing services; her attention to detail made the editing process both simple and straightforward. Thanks are also due to Jennifer Taylor, also of Old Dominion University, for her kind assistance and expertise in formatting the document. Finally, we thank our wives, Donna Kay Harrison and Elizabeth Dashiell (Betsy) Morris; their limitless love and support ultimately made this book possible.

Introduction

John C. Morris and Gerald Andrews Emison

Environmental policy has developed into a complex undertaking in contemporary America. For the most part our society has settled the argument about whether this is a valid public and private undertaking. The answer is "Yes, but how shall we do so effectively, efficiently, and fairly?" This book seeks to explore this question for the South, an important and distinct region of the country. By doing this, we hope to shed light on this question for the rest of the nation.

The complexity of environmental policy today extends in a number of directions. First, it is complex in analysis. A single dimensional analysis will not suffice for us to understand the policy situation. We must consider simultaneously the political, economic, geographic, managerial, and social dimensions of environmental protection. As a result, there is no mechanistic toolbox for us to deploy in analyzing environmental policy. We must consider multiple dimensions concurrently, and there is no way to do this other than the old fashioned approach of reasoning simultaneously in multiple dimensions. Herein we seek to present environmental issues confronted in the South in order to advance such simultaneous, multidisciplined approaches.

Environmental policy is complex not only in analysis but in implementation, and this complexity of implementation adds further impetus to understanding the particulars of environmental policy implementation in a specific setting. The interaction of policy design, political interests, different governmental capacities with private sector capabilities and motivations creates a yeasty setting in which intentions often are overtaken by unexpected realities. Understanding the particulars of such influences can equip us to both understand and act more effectively when other unique circumstances arise.

This complexity accelerates more when complexity in analysis and complexity in implementation interact in a specific geographic setting. In short, geographic variability is important. Examining such variability in one region can yield insights into environmental policy implementation in other regions.

This book examines the South because how this distinctive region faces environmental issues is important, not only to the South, but to the rest of the nation. We undertook this volume to learn about the fundamental features of environmental management in order to advance insight that may be of use across regions. We believe that the actual circumstances surrounding implementing environmental policy in the South can yield practical knowledge that can improve the practice of environmental protection regardless of national region. We also undertook this book because we believe that the interaction of fundamental features of policy with the unique features associated with a region can improve our understanding of how complex environmental policies can be improved.

There are many different definitions of "The South." One common definition is to include the members of the nineteenth-century Confederate States of America: Alabama, Arkansas, Florida, Georgia, Louisiana, Mississippi, North Carolina, South Carolina, Tennessee, and Texas (see, for example, Black and Black 1987; 1990). Other work (Breaux, Morris, and Travis 2007) omits Texas from the definition, but retains the same "core" states. One might also argue that Kentucky and West Virginia are reasonably included in the region.

For our purposes, we define the South as the contiguous region encompassing the states of Virginia, North Carolina, South Carolina, Georgia, Florida, Alabama, Tennessee, Mississippi, Arkansas, and Louisiana. In addition to their inclusion in most definitions of the South, they share in common another important trait: they tend to be dominated by traditionalistic political culture (Elazar 1984). V. O. Key (1949) was among the first to formally acknowledge southern "distinctiveness," and while he found significant political differences both within and between states in the region, he made specific note of the characteristics that made the region distinct from the rest of the nation. Citizen attitudes, ideology, resources, religious traditions, politics, history, and language all interact to make the South distinct from other parts of the country; we argue that these factors are largely shared among the states in the region. In short, we proceed from the premise that southern distinctiveness is both interesting and important in the context of environmental policy implementation and management.

There are a number of themes that appear in the following chapters regardless of pollution medium or institutional setting. The South offers an ongoing clash between the preference for traditional roles and responses of governments, business, and the public, and the persistent requirement to innovate to protect the environment effectively. This simultaneity of conventional methods and innovation yields noteworthy conditions that are of interest not only in the South but across the nation.

The South also is of interest because it offers the full range of environmental issues. Conventional pollution control through installation of control devices, conflicts over allocation of natural resources such as water supply, and issues of hazardous site cleanup occur in the region alongside challenges of environmental justice. While the South is unique, it also is representative of a number of environmental issues that are important regardless of location in the nation.

Because the South presents this complex, full range of issues, the chapters that follow employ a range of methods to explore specific environmental challenges. The stresses in the South of growth, investment in pollution management, and changing roles of public and private institutions, as well as the emerging challenges that come from understanding the interaction of physical conditions, economics, and politics, offer ample opportunity to explore this region and its environmental issues through a variety of methods matched to the unique environmental circumstances. While the chapters in this book do not cover all of the states in the region, we choose them because they are representative of the challenges and solutions found in other states in the region. If we are correct in our assertion that these states share common characteristics (that are in turn somehow different than other states), then the states and cases selected for this volume should aid our understanding of similar issues in the other southern states.

Chapter 1 examines the emerging conditions that are propelling change in environmental policy and management. This chapter introduces the idea that postmodern environmental management is emerging toward systems that simultaneously build upon traditional approaches while transforming themselves through adaptation. This chapter frames the value of exploring the South to understand this emergence and sets the stage for the subsequent case studies that follow.

In chapter 2 Breaux et al. examine state commitment to environmental quality in the South. Their examination of a range of potential factors influencing various measures of intensity of commitment shows the importance of interest groups, pollution levels, and state wealth. But this work raises questions about per capita spending on environmental matters as influential on environmental quality. In doing so the authors establish the complexity associated with environmental commitment across the South and identify the clash of tradition and innovation.

In chapter 3 Gerald Andrews Emison analyzes progress on ozone air pollution in the South and finds that the South is distinct from other regions of the nation. This distinctiveness is associated with low resources directed to overall environmental protection rather than specifically with air quality. In

doing so Emison concludes that narrow, pollution medium-specific allocation of resources may be inadequate to improve air quality. A more general
state capability at the environmental infrastructure level seems important.
Such a finding suggests a need in the South to build overall environmental
capacity as a strategy to address complex environmental problems.

John Morris examines in chapter 4 three southern states' responses to federally required capital investment assistance programs for clean water and
for drinking water. His findings suggest that in response to similar federal
requirements, states develop programs that incorporate their unique policy
perspectives and institutional profiles. In doing so Morris draws attention
to the underlying property of variability of adaptation to similar evolutionary stresses. This identifies the importance of tailored analysis that reflects
similarities and differences among states that are analyzed for similar environmental policies.

The importance of state-by-state variability of responses to similar pressures appears in chapter 5, James Newman's case of Alabama's and Georgia's
behaviors in water supply negotiations. Each state relied on its own unique
political culture as they interacted with each other. Also, Newman finds a
simultaneity of responses as the states negotiated with each other over water
supply matters of surpassing concern to each one. Evidence of the interaction
of tradition and innovation characterized these states responses.

Even when the pollution medium moves from water to hazardous waste
cleanup, Deborah Gallagher shows in chapter 6 that southern states seek
innovative approaches while building upon established approaches. Her
examination of North Carolina's and Florida's approaches to employ
public-private partnerships demonstrates the preference for holding to important traditions yet striking out in policy directions as yet well-defined. In
addition Gallagher draws attention again to a variability of state responses
to similar environmental pressures.

In chapter 7 Celeste Murphy-Greene examines responses across the South
to federal Pesticide Worker Protection Standards. Her analysis finds violations of these standards to be widespread across the South. In doing so
she demonstrates the persistence of timidity of southern states' responses
to an emerging environmental issue. This documentation of adherence to
traditional approaches of caution that may lead to underprotection of at-
risk groups confirms the strength of tradition among some environmental
management efforts in the South.

The reaction to change by one southern state's environmental agency, the
Mississippi Department of Environmental Quality, is explored by Gerald
Andrews Emison in chapter 8. In doing so Emison identifies the presence of
unconventional responses to new challenges by the state agency. But these
responses are limited by adherence to tradition in behaviors by a range of
actors who influence the agency's behavior. In short Emison presents a case
in which both adherence to traditional approaches and innovation occur

simultaneously. He examines the conflicts this creates and the implications for responses by the agency.

Collaboration as a response to environmental problems is a central theme of Madeleine McNamara's study of the Virginia Coastal Zone Management Program, presented in chapter 9. While the successful use of collaborative structures in a traditionalistic political culture is often considered to be difficult, McNamara's case study not only explains why their use is possible, but suggests that many of the traits of successful collaborations are fully compatible with traditionalistic cultures. Collaboration can thus be an effective mechanism to help meet environmental goals when existing structures and resources are inadequate.

Erin Holmes analyzes in chapter 10 the conditions of a newly emerging environmental issue in the South, expansion of the biofuels industry. With agriculture a prominent economic and political feature of the South, she examines the implications this has across the region. Cohesive responses are still developing, yet she finds the consistent effect that political alignment, political influence, and geographic similarity have on emergence of the biofuels industry.

The book concludes with chapter 11. This chapter draws together the threads from the individual chapters. In doing so the chapter identifies the importance of context, whether geographic, political, or economic, in understanding responses to similar environmental issues. In the South we find a preference for adherence to traditional roles and responses, nevertheless we also find state actions that seek to respond to emerging environmental issues. These responses have unique properties, however. They seek to stay within traditional patterns of responsibilities. They evince willingness to innovate, but this is a constrained innovation, and we find that this constrained innovation yields outcomes that have very uneven results. In the South today we find a willingness to change, but as Key found over fifty years ago, an adherence to tradition still constrains policy responses.

REFERENCES

Black, M., and E. Black. 1987. *Politics and Society in the South*. Cambridge: Harvard University Press.
———. 1990. The South in the Senate: Changing Patterns of Representation on Committees. In Steed, R. P., L. W. Moreland, and T. A. Baker, eds. *The Disappearing South? Studies in Regional Change and Continuity*. Tuscaloosa, AL: University of Alabama Press, 5–20.
Breaux, D. A., J. C. Morris, and R. Travis. 2007. Explaining Welfare Benefits in the South: A Regional Analysis. *American Review of Politics*, 28, 1:1–18.
Elazar, D. J. 1984. *American Federalism: A View from the States*, 3rd ed. New York: Harper and Row.
Key, V. O. 1949. *Southern Politics in State and Nation*. New York: Knopf.

1

Tradition and Adaptation

The Postmodern Environmental Management System

Gerald Andrews Emison and John C. Morris

INTRODUCTION

During the 1970s and the 1980s the United States substantially improved its environmental quality. It achieved this through the command and control regulatory system. This system is the exercise of the coercive power of the state to compel action by establishing standards and enforcing those standards through administrative and judicial actions. Such an approach depends on centralized knowledge, application of authority, and limited participation in decisions by those who must carry out such decisions (Rosenbaum 2005). Since the 1990s national environmental management has experienced a number of pressures that call into question continued use of this conventional approach (Fiorino 2006). The nation has begun to explore variety in environmental management in order to identify new, effective models. State governments appear as an integral category of institutions under any new environmental management approach (Rabe 2006). Successfully engaging these changing conditions requires exploration of a range of approaches, and the following chapters examine experiences of southern states as they have confronted this changing world of environmental management.

At the heart of such variation is the capacity of the states to implement and manage national environmental policies and standards. We begin from the premise that states are inherently dissimilar in their abilities and (political) interests when it comes to the implementation of federal policy objectives, a position that becomes even more apparent when one considers the substantial complexity of both environmental problems and their chosen solutions. It is a widely accepted observation that environmental problems do not respect political boundaries, which means that any truly effective policy aimed at environmental protection must be national in scope. At the same time, since policy implementation is largely carried out at the state level, common sense suggests that variation in the abilities of the states to carry out such national policies also needs attention.

The purpose of this collection of essays is to more fully explore the means by which southern states address environmental imperatives. A diverse literature developed over several decades has largely concluded that southern states tend to lag behind the rest of the nation in terms of a willingness to address new policy initiatives. Moreover, conventional wisdom holds that southern states are, if not incapable of, at least uninterested in innovation. The reasons for this are many and diverse, and span the gamut from issues of resources to political will to administrative capacity. Yet southerners have traditionally espoused a strong connection to their environment, a connection that runs deep in southern culture. Furthermore, southern political culture tends to be largely antigovernment, and southerners often see government as a negative force (see Elazar 1984). At the same time, environmental quality gains are most often the result of government action, a statement as accurate for southern states as it is for the rest of the nation.

Despite a more common belief that southern states are laggards when it comes to policy innovation, recent evidence suggests that southern states are, in fact, quite innovative (Breaux et al. 2002; Breaux, Morris, and Travis 2007). The key to understanding the policy behavior is to understand the context in which policy makers and policy implementers operate. V. O. Key's (1949) seminal work on southern politics suggested that southern politics were truly unique, and largely the result of the particular history and traditions of the South (Key 1949). This scenario raises an interesting question: Can southern states innovate in the area of environmental policy, and if so, does this innovation take unique forms?

The chapters in this book illustrate ways in which southern states have been innovative in both environmental policy and environmental protection. By highlighting the ways in which policy makers in the South have sought to address environmental issues within the constraints of the larger political and social context, we can better understand not only the policy outcomes that result, but why certain outcomes occur—both positive and negative. If politics is truly the "art of the possible" (Von Bismark 1869), then context defines the boundaries of the "possible."

A BRIEF HISTORY OF ENVIRONMENTAL POLICY AND MANAGEMENT IN THE UNITED STATES

Although most Americans might suggest that federal environmental policy dates back to the 1970s, the national government has been involved in environmental policy for more than a hundred years. The Refuse Act, passed in 1899, was an attempt to protect the navigable waters of the United States by making it illegal to dump refuse into these waterways (Freeman 1990). Although the purpose of the legislation was to prevent hazards to navigation rather than to ensure clean water, it had the effect of protecting water quality by stopping the wholesale dumping of sludge, sawdust, and other materials into these waterways. Other early landmark federal legislation included the Water Pollution Control Act of 1948 (Freeman 1990) and the Air Pollution Control Act of 1955 (Portney 1990).

During the 1960s, several well-publicized events began to galvanize both interest groups and the general public to call for greater federal involvement in environmental protection. Rachel Carson's (1962) seminal work, *Silent Spring*, called attention to the ecological harm caused by the use of some insecticides. Increasing air pollution problems in the Los Angeles basin and around Houston, Texas, and Birmingham, Alabama, began to raise awareness of air pollution problems. A significant oil spill off the coast of Southern California made headlines for weeks, and television footage of a burning Cuyahoga River in Ohio served to generate calls for water pollution controls. At the same time, scientific studies began to detail the deleterious effects of different pollution media on the health of the environment. The calls were clear enough that Richard Nixon made the expansion of federal environmental policy an important part of his election platform, and shortly after taking office in 1969 he proposed the creation of the Environmental Protection Agency (EPA) as an independent executive branch agency.

The creation of the EPA also served as a catalyst for a string of new federal environmental legislation in the 1970s. The Clean Air Act of 1970, itself a revision to the Air Quality Act of 1967, required states to undertake very specific actions to address air quality, and to reach a set of goals in terms of air quality standards (Portney 1990). Two years later Congress passed the Federal Water Pollution Control Act Amendments of 1972 (more commonly known as the Clean Water Act), which provided, among other things, significant resources in the form of categorical grants to help communities build or expand wastewater treatment facilities (Heilman and Johnson 1992). The Resource Conservation and Recovery Act of 1976, followed shortly by the Comprehensive Environmental Response, Compensation, and Liability Act (CERCLA, better known as Superfund) in 1978, were designed to address problems of hazardous wastes (Dower 1990). The Toxic Substances Control Act of 1976 addressed problems related to a broad range of toxic chemicals in the environment (Shapiro 1990). These congressional mandates, along with a host of others, greatly increased the federal role in environmental protection.

Nearly all of the above-mentioned legislation contained a common theme: The national government took the lead in providing goals, resources, and expertise to address these environmental problems. States played a role in the implementation of these legislative requirements, but the model was very much a top-down command and control model. By the mid-1980s, however, the broader approach to federal environmental protection had shifted yet again. The "Reagan Revolution" of the early 1980s fundamentally redefined the role of the national government and the mechanisms through which national policy would be implemented. Driven by a strong desire for a smaller, less intrusive national government and an expansion of states' rights, the Reagan administration began to search for policy solutions that would achieve these political goals, while simultaneously reducing the financial burden on the national government (Morris 1997). The first major piece of environmental legislation enacted that reflected these new goals was the Water Quality Act of 1987. Among other things, the act largely did away with the use of categorical (construction) grants for wastewater needs, and instead capitalized state-administered funds from which states would serve the needs of communities within their jurisdiction (Morris 1996). The penalties for noncompliance were still in place, but the program fundamentally redefined the federal approach to environmental protection and its concomitant regulatory approach.

Other significant changes took place in the 1990s and into the turn of the century. The Clean Air act of 1990 was a complex, long (more than 750 page) (Portney 1990) document that introduced market forces into environmental protection. Congress passed the Safe Drinking Water Act in 1996, but the act was so complex that it took several years to produce the required regulatory guidance for the act. The Clinton administration attempted to "fast-track" publication of the rules, but failed to complete the process before Clinton left office in early 2001. The Bush administration let the matter drop.

In sum, there has been a significant increase in the federal role in environmental protection since the early 1970s, as the national government moved to address demands made by citizens to protect the environment. The resulting policies were largely implemented, reflecting the severity of the perceived problem, if the not the strength of the demands for action. Yet all of the early policy instruments contained a common component: They were all reliant on strong command-and-control mechanisms for their management and implementation, and the roles for the states were generally subservient to those of the national government.

THE IMPLICATIONS OF COMMAND AND CONTROL ENVIRONMENTAL MANAGEMENT

During the 1970s and 1980s the nation rushed its environment to the emergency room. The U.S. environment was suffering a decline in vital signs due to long-term abuse and neglect. In an emergency room, preservation rather than enhancement is the watchword, and the techniques applied are expedient rather than precise. That was indeed the case with the U.S. environment in the 1970s and 1980s—fairly uniform command and control regulations were administered in massive doses to stabilize the patient. They worked: the patient's vital signs stopped their decline and, in fact, improved (Rosenbaum 2005). But patients do not recover in the emergency room. At best the ER only prepares patients to take full advantage of the recuperative phase that follows. In fact, if patients are kept in the ER too long, they become dependent on the life support and do not develop their own capacities to return to a healthy life. A closer look at our command and control response to environmental problems suggests a similar liability for using it for long-term environmental quality efforts.

In an emergency mode treatment is prescribed by experts. The emergency room intrinsically depends heavily upon care externally applied to the patient. In the ER the patient is not expected to participate actively in the intervention. The protection measures are determined by specialists without much assistance from the patient except in describing the symptoms. Conventional environmental regulatory approaches have used either legislative or administrative prescriptions, almost always relying on technology that has been prescribed by experts. The affected stakeholders have played limited roles in designing the corrective measures. In other words, conventional environmental management has not depended much on the active decisions of industry, state and local governments, or citizens in designing actions. It has expected that these parties will "take their medicine and do what the doctor orders." For example, the requirements for alternative fuels in the Clean Air Act represent such a prescription. They were developed by experts and imposed on the stakeholders. The alternative fuels program was premised upon such externally chosen and applied intervention (Bryner 1995).

Treatment in the emergency room often depends upon invasive intervention; traditional command and control regulations similarly use invasive measures. Cardiac patients brought to an emergency room often face application of medicine to stabilize their heart rhythms. These medications are necessary to prevent immediate catastrophe, but do not work well as a long-term healing strategy. Command and control regulations often specify precise operating behavior and capital investments by polluting sources. Whether controlling emissions from floating roof storage tanks or treatment of hazardous wastes, the command and control regulatory system depends upon the government's highly precise specification of operations and investments for success (Lentz and Leyden 1996). Little latitude is afforded those who wish to vary from the regulations, even if alternative means to achieve the desired results are possible. This intrusion into operations was necessary when sources of pollution lacked the motivation or skill to perform (Spence 1995). When motivation and skill improve, sticking with such an invasive approach can limit performance: It takes away flexibility to find a better solution than the prescribed one and sets a firm ceiling on performance outcomes (Fiorino 2006).

Command and control regulations share another trait with emergency rooms: The interventions are usually well-tested and carefully focused. Experimentation is confined to that which is essential to validate a treatment's adequacy, not optimality. The actions can be risky; indeed, much is at risk and timidity and caution can result in catastrophe. By sticking with an established protocol, unforeseen complications can be minimized. Gains from innovations may be foregone, but survival, not enhancement, is the objective. When the nation's environmental quality was declining precipitously in the 1970s, the prescription of command and control regulations could be justified by the seriousness of the situation. The need to see the command and control regulations through to achieve major improvements was necessary to avoid confusion at a crucial time. Since environmental conditions have improved, we can broaden our attention to include long-term environmental enhancement. It may make sense to shift to pollution prevention rather than continue to focus on increasingly restrictive end-of-the-pipe controls.

Both emergency rooms and command and control regulations are premised on remedial measures. Both seek to correct something gone very wrong. The ER often seeks to remediate trauma; command and control regulations seek to halt decline in environmental quality by abating widespread emissions into the air, land, and water. Both aim to correct rather than enhance. A patient has little hope for the long-term unless the treatment objective shifts from remediation of the trauma to improvement. Similarly, we may staunch emissions to our air, land, and water, but unless we enlist the efforts of industries, interest groups, and consumers to take direct action to improve, not just to protect, the environment, we face a future of fighting against a long-term retreat from a healthy environment. A nonprofit organization's land purchase program may protect and improve larger ecosystems more effectively than any continued application of the Clean Water Act's Section 404 dredge and fill program (Schmidheiny and Zorraquin 1996).

A maxim of emergency room care is to limit the reliance on extreme but necessary measures. Patients can readily lose their ability to breathe independently if they are kept on a respirator longer than necessary. Everything that is necessary in the ER is done to preserve life, but no more is done than is necessary. Command and control regulations provide powerful medicine for halting environmental decline. But when these regulations are administered beyond the necessary horizon, they can promote dependency and limit innovation. This prevents development of longer-term, self-sustaining patterns of response that offer more than the hope of damage remediation. They offer the potential for environmental enhancement.

In other words, like the ER, command and control regulations protect during traumatic times. But a life lived simply for the purpose of surviving trauma is not likely to last or to be of high quality. The recuperative phase depends on a patient developing independent means of living. Similarly, as long as the objective for environmental quality remains damage mitigation or simple protection rather than enhancement, the chance for a robust, sustaining environmental policy is low. It is very difficult to be successful only by avoiding catastrophe (Behn 2006). Advancement and improvement as policy objectives offer the best chance for developing a rich environmental quality as part of our national fabric. The policy question environmental management faces today concerns how to adapt to such a new regulatory approach (Fiorino 2006).

Yet today we are close to a stall-out in environmental quality improvement. The nation's political leaders seem determined to draw battle lines over environmental protection by debating the degree of regulation desirable (Kraft 2006). This debate has evolved into something of a ritual. Today's debates sound distressingly familiar to the debates of the early 1980s. Discourse concerning environmental quality centers around budget allocation decisions and how much regulation is necessary. The entire public discussion concerning environmental quality is taking place in a flat policy plane where the degree of regulation and the amount of public funds allocated to environmental protection are posited as the only dimensions. This debate is conventional, and it prevents the nation from considering more innovative approaches. The opportunity cost of this debate is forgone innovative approaches and, by extension, the chance for an enhanced environment.

Rather than discussing environmental policy in the plane of regulation and expenditures, the nation needs a more realistic landscape. This landscape, rather than being flat, has multiple dimensions. Topics such as the degree and direction of scientific research, the amount and nature of technical assistance and training, the role of enforcement, the role of geographic information, environmental education, and pollution prevention serve notice that an environmental policy limited to a flat policy plane is inadequate. With an inadequate policy framework, the nation risks overlooking possible gains by oversimplifying candidate policies. Today's public discussion offers little beyond the traditional command and control approach. Most environmental policy prescriptions stop with repairing damage. If we learn from the clinical model of emergency room recovery, we also need to seek policies that prevent a return to the ER.

Until now, national environmental policy has focused on "protection." As Wiener (1995) observes, most of today's theory of environmental quality efforts depend on the view that the environment must in some way be "preserved." This derives from the belief that the environment has some stable, pristine condition that would exist but for the presence of humans. Wiener notes that recent insights into ecology demonstrate that such a static view is at variance with facts. The natural systems that are the subject of environmental policy are constantly changing and adapting. Any policy built on a view of the environment as unchanging is likely to fall short of its stated objectives: No policy can hope to succeed if it contradicts fundamental principles, no matter how deeply embedded in conventional wisdom and values it may be. If humans are part of the changing environment, a key question becomes how humans can persistently adapt their environmental policies to be in concord with the changing physical environment (Wiener 1995).

Rather than establishing protection as an environmental goal, Fiorino (2006) suggests we consider environmental enhancement and improvement as a national goal (Fiorino 2006). The physical environment restlessly seeks progress and improvement. Our national policies could seek to reinforce such tendencies and make progress a central feature. A constant search for adaptation and improvement is inherent in systems whether ecological, biological, or political (Kaufman 1995). We have built artificial walls around both the concept of environment and the advancement of environmental goals. If we look beyond conventional approaches perhaps we may find ways to enhance, rather than simply protect, the environment. This book seeks to do this for one region of our country and extract lessons that other areas may find useful.

EMERGING ENVIRONMENTAL CHALLENGES

Discovery of novel approaches is essential to sustaining environmental progress into the future of changing circumstances. These novel approaches are essential because the conditions of environmental quality are changing. What evidence indicates it is time to reconsider the basis of national environmental policy?

The nature of pollution is changing. Pollution from large industrial sources characterized much of the nation's previous environmental challenge. Control of large point sources was appropriate since those large sources were producing most of the pollution. Today most of these sources are under control or on compliance programs, and dispersed sources cause a large portion of our pollution problems. Many of the nation's air pollution sources are small sources such as dry cleaners, paint shops, construction sites, and the ultimate small source, automobiles. Today almost 60 percent of the volatile organic compound emissions into the nation's air come from nonindustrial sources (U.S. EPA Office of Air Quality Planning and Standards 2005). Nonpoint sources of water pollution have long been identified as leading sources of stream pollution (U.S. EPA Office of Water 1992). These sources are not only smaller than before, they are more extensive, spread throughout our urban areas rather than concentrated in a few large places. Whether it is urban runoff or service station degreasers, our source categories have changed significantly. Further, since these sources are so small, they often are very competitive with each other and are consequently exquisitely sensitive to costs such as those imposed by command and control environmental regulation. We find that whether it is ozone in our cities or stream quality in our rural areas, small, dispersed sources comprise much of the current problem. These are problems at a different scale that call for different skills, approaches, and tools.

There are new remedies that must match the new pollution. With the focus shifting from large point sources to many dispersed sources, we can no longer rely on "bolt-on" control devices to abate pollution. We are looking at employing best management/operation practices and preventing pollution before it is generated to combat unwanted discharges (Gunningham and Grabosky 1998). Regulation based on static production processes is doomed to obsolescence before it begins. Industrial sources today tend to adapt quickly because they are flexible and highly competitive. Many industries change their production lines daily. Command and control regulations that target processes rather than emissions often chase perpetually moving targets. This implies turning to different tools if we are to continue to make progress.

The more we deepen our understanding about the character of pollution, the more concerned we have become. As scientific knowledge explores the consequences of pollution, we find effects we were unaware of ten years ago and at concentration levels previously unknown. In the 1970s we were just learning of the serious health effects of particulate matter in our lungs. Today we regulate even fine particles (U.S. EPA Office of Air Quality Planning and Standards 2005). We are also learning the degree of interconnectedness of our ecosystems. Impacts on biological diversity and species survival seem to grow as we increase our knowledge (Dietrich 1995). Ozone layer depletion was not clear until the late 1980s (U.S. Council on Environmental Quality 1993) . As we learn more, we see that the consequences of humanity's use of technology extend farther than we have ever imagined. The current debate over global climate change stands as the most recent memorial to the evolution of environmental knowledge.

When we consider the processes in use today, we find that both public and private management are emphasizing the value of sharply drawn goals with emphasis directed to conditioning organizations for success (Behn

2006). As a result, more diverse responses are tailored to the unique circumstances of an emission situation.

The resource landscape for environmental protection is growing and contracting at the same time. Sharp limitations on funds available at the federal level for environmental activities have characterized the 1990s and 2000s (Rosenbaum 2005). While the allocation of funds has grown, given the size of the federal deficit, it is very unlikely that we will see resources rising to meet expanding responsibilities at the federal level in the foreseeable future. A proverb states, "We have no money, therefore we must think." Our national environmental quality program will need to rely on creativity to replace resources.

The institutional setting for environmental protection is shifting and government relations are changing. State and local governments have paradoxically developed greater strength than ever to meet environmental challenges while experiencing retrenchment in their basic capacities due to fiscal limitations. These governments have more legislation, more technical and managerial skills, and more resources than ever to take on environmental problems. However, they are being squeezed by the same promise/performance gap that the EPA faces. In many instances, they face even more severe budget limitations because state and local revenue bases do not keep pace with required expenditures (Ringquist 1993; Scheberle 1997; Rabe 2006).

Among our business institutions, corporate thinking about the environment has shifted in many companies. Many firms seek to enhance their market presence by emphasizing environmental features. In such cases the environment is not a constraint but a potent competitive advantage. "Green" goods and "green" behavior are increasingly seen as a way for responsible corporations to bolster their bottom line. Corporations are moving as never before to include the environment in the basis of doing business (Breen and Anastas 1994; Schmidheiny and Zorraquin 1996). A major policy challenge is how to capitalize on this and enlist the talents of corporate self-interest to advance environmental quality.

While the circumstances have changed, reliance on the command and control model has not. This has produced a climate in which public actions induce discord, not progress. This discord occurs between governments, such as the federal and the state levels. News media report ongoing antagonism between the EPA and state pollution agencies (Rosenbaum 2005). Whether the issue concerns use of alternative fuels for automobiles, inspection/maintenance programs for smog controls on automobiles, or national standards to address global climate change, the range of conflict is large. Quarles (1995, 3) contended that "careful design of environmental regulation has been obstructed by intense emotional polarization. . . . When tempers flare and debates degenerate to shouting matches, a thoughtful analysis of complex questions becomes almost impossible." Such discord, whether between governments or between the public sector and the private sector makes it difficult, if not impossible, to seek common solutions. Preferred solutions would depend on win-win strategies rather than win-lose attitudes, but today's discord pushes stakeholders into a win-lose mindset. We do not systematically search for new means to pursue national environmental goals; all sides look for means to protect narrow short-term interests (Quarles 1995).

A reading of the national environmental statutes shows a strong legislative statement in favor of the environment. Congress has consistently stated its policy objectives to be to "protect and enhance" the nation's environment. But an equally careful examination of the means of achieving this shows "protection" to be the dominant approach. The nation has, rather than advancing environmental goals, sought to limit the degradation of the environment to acceptable levels. This places environmental quality as a secondary goal of the country. It is something to be used as a constraint; it has not been given equal legitimacy with other social objectives that the polity seeks to maximize. In other words, we have as a basic assumption that policy should be governed by containment of damage rather than improvement of the physical environment. As Spence (1995) has pointed out, environmental quality has reached such a deeply rooted point in American values that it pervades the political value system. No longer is it necessary to confine public policy to damage avoidance; public opinion sees environmental quality as intrinsically valuable and, hence, should be the subject of continual efforts to improve it. Yet our policy system does not presently reflect that change.

Morris asserts that the hallmark of success is constancy of goals and flexibility of methods to attain those goals (Morris 1994). Failure, according to Morris, often contains unwavering commitment to methods while goals are changed to ones that are attainable through such familiar methods. If a healthy environment remains the national goal, it may be time to reconsider the methods. Reliance on protection, as contrasted with improvement, and reliance on command and control may be outdated as single-purpose approaches. It seems we need approaches that have multiple methods and do not depend upon the right action of any single player alone. We must match problems, responses, and institutions, and do so in a manner that attends to today's public values of competing interests. Transparency of methods and goals coupled with policies that are so agile as to permit easy shedding of problems once they are generally corrected require approaches that foster discovery and innovation so that we incorporate new knowledge easily. In short we need a new way of thinking of environmental policy.

ENVIRONMENTAL POLICY AND MANAGEMENT FROM THE PERSPECTIVE OF THE STATES

An expanded role for state governments in the implementation of national environmental policy brings into focus the ability and willingness of states to carry out the necessary policy actions. If states are truly "policy laboratories," as Dror (1968) contends, then it follows that states will take different approaches to meet the national policy goals. The state comparative literature is rife with work that details the many factors that impact state policy choices, but the underlying premise that states do make choices is undisputed. In short, state differences matter, and they matter a lot. Moreover, much of the existing work focuses on a national comparative structure, in which all fifty states are addressed simultaneously. While such an approach tells us a lot about the national picture, it also tends to obscure important regional differences that may be present.

The southern states are especially worthwhile to examine in such a manner. Although much has been written about the politics of the South since the publication of V. O. Key's seminal work in 1949, there has been little focus on environmental policy and management in southern states. Yet if the characteristics of the South are as unique as the broader literature suggests, an examination of these policy activities in this setting seems all the more worthwhile. The cases discussed in this book demonstrate the effort, conscious or inadvertent, to innovate while adhering to effective environmental protection simultaneously. With this comes the acknowledgment that sometimes experiments fail. How the southern state governments do this within their unique context provides important lessons for facing future environmental management challenges at the frontiers of state environmental management.

The collection of essays in this volume provide a framework through which we can begin to understand the choices states make when it comes to environmental management and implementation. Each of the chapters addresses a different pollution medium or environmental issue, and the forces behind each, and their explanators, can differ significantly. Rather than impose an artificial framework through which to analyze these different issues, we have encouraged the authors of the respective chapters to choose the framework that best illustrates the forces at work in that realm.

While the themes in each chapter thus differ somewhat, there are several themes that are common across all chapters. Each theme is reflected, to different degrees, in each of the following chapters. These themes are not meant to be rigid, but rather to serve as unifying threads through which environmental politics and management might be tied together. These themes are discussed again in chapter 10.

First, politics matter. The ways in which political decisions are made vary from state to state, and the outcomes of that process also matter. In some cases, this can be explained by determining which political party controls the state legislature or the office of the governor; in other cases, legislatures and governors do things seemingly at odds with the stated positions of their political parties. Likewise, some legislatures are collegial, while others are very contentious. The constitutional powers of the governor can also play a role—some governors have weak powers, while others are granted significantly more policy authority.

Moreover, the terms under which politics are conducted within a state can matter greatly. For states in which there is a high degree of consensus, or conversely, a lack of partisanship, we are likely to detect different policy outcomes than states in which the reverse is true. The greater the degree of consensus, the more likely the policy instruments produced in that environment will be free of ambiguity and conflicting goals. The conduct of politics also has significant implications for the implementation and administration of environmental policies. A higher degree of political consensus is likely to produce clearer instructions to implementing agencies, and thus less delay in implementation.

Second, the adherence to tradition is an important characteristic of southern culture (Key 1949). The centrality of tradition, and its concomitant conservative approach to policy, helps define state approaches to policy. The political culture of the southern states tends to be more consistent across the states in the region than for other parts of the nation. Elazar's contention that the historical migration of different religious groups helps define today's political culture in the states is a useful, if somewhat ill-defined, theme. While the dominant traditionalistic culture of the southern states seems to limit the kinds of action possible in the management and administration of environmental policy, we find strong evidence of a "counterculture" present in terms of the ways in which state environmental agencies seek to implement policy. This is especially true at the margins— while Mississippi is the most "traditionalistic" state in the nation, North Carolina is characterized as traditionalistic/moralistic; Florida is traditionalistic/individualistic, and so on. The point here is that while the political culture of the states is very similar, it can differ greatly at the margins, and these differences matter in terms of the ways in which states approach environmental policy and management.

Third, resources matter. By their very nature, environmental policies often require significant amounts of resources to implement and administer. Resources can take many forms, and all are important in policy terms. Political resources can include the political will of the legislature and the governor, but also includes the existence of demands for policy action by both individual citizens and organized interest groups. The South is generally considered to be "business friendly," and such a designation often suggests that environmental interests are generally powerless. While this is generally true in the South, the balance between business and environmental interests can vary from state to state, and even from issue to issue (note the designation of North Carolina in chapter 7 as the birthplace of the environmental justice movement). Monetary resources are no less important, and smaller and less powerful economies of the southern states means they generally have fewer resources to commit to environmental needs. Even so, states can differ significantly on this dimension—Florida and Mississippi are on opposite ends of the resource scale. Finally, administrative resources are also important. Although most state governments are more professional today than at any other time in our history, significant differences between states still exist; these differences have implications for a state's ability to assume responsibility for the implementation and management of national policy goals.

Fourth, the pollution medium matters. Where the environmental problems are well understood, where the scientific community is in agreement, and where the institutional structures have developed over time, innovation in environmental policy and management is both evident and largely embraced. On the other hand, where the science is less well understood, or where institutional structures are relatively new or absent, innovation is often constrained. Likewise, the ability of humans to address and ameliorate environmental damage caused by a specific pollution medium also plays a role.

Fifth, innovation is best understood in the context of the individual state. If Dror (1968) is correct in his observation that states are "policy laboratories," and Key (1949) is accurate in his description of the political and cultural differences between the southern states, then we should reasonably expect state contexts to differ. An implication of these differences is that choices available to one state may not be available to another. If choices are constrained by context, then a routine decision in one state may well be an innovative solution in another. Likewise, one person's "common-sense solution" may be another person's innovation. In the context of welfare reform, for example, Breaux et al. (1998) note that Mississippi, not known nationwide as a particularly innovative state, actually developed a waiver program under the Aid for Families with Dependent Children (AFDC) program that became the model for AFDC's replacement policy, Temporary Aid for Needy Families (TANF). Mississippi did not consider their choices under AFDC to be innovative; it was just "good policy."

Finally, the world of environmental policy is complex. First, the policy settings in which environmental policy is formulated and implemented are highly diverse. Because environmental problems do not respect political boundaries, both the policy problem and its solutions require cooperation between different groups of policy makers. While federal legislation seeks to impose requirements on states, recent trends in environmental policy seek to give states greater discretion in how they meet these requirements (see Morris 1997), and how they structure both their policies and their implementation apparatus. In this sense, states truly become policy laboratories, and how the states respond in this setting become worthy of study. Second, because the policy problem (pollution) is complex, the solutions to address the problem are also often complex. The environmental policy arena is largely driven by scientific evidence and debate, which has the effect of moving the debate beyond the knowledge and comprehension of the average state policy maker. This state of affairs places additional authority and influence in the hands of unelected officials. In addition, complex problems often result in implementation structures that attempt to bring to bear as many resources as possible to ameliorate the problem. This often translates into implementation structures that may involve many governmental agencies spread across several jurisdictions. Managing such complex structures successfully can be daunting task.

DEFINING "GREEN" IN THE SOUTHERN CONTEXT

The following chapters in this volume illustrate the ways in which southern states address environmental imperatives in their states, how they meet (or fail to meet) expectations for environmental quality, and the factors that help determine their successes or failures. Are southern states destined to serve as poorly equipped "emergency rooms" for environmental imperatives? Are the states able to embrace postmodern environmental management, or are they doomed to remain trapped in a twentieth-century bureaucratic paradigm? Is command-and-control environmental management appropriate for the contexts of these states, or is an alternative approach preferable (or possible)? While it is unlikely that any single volume can answer these questions definitively, we can begin to define the terms under which the questions can be both asked and addressed.

REFERENCES

Behn, R. 2006. Avoiding All Mistakes. *Public Management Report* 4, 4 (December), www.hks.harvard.edu/thebehnreport/December2006.pdf (accessed February 6, 2010).

Breaux, D. A., C. M. Duncan, C. D. Keller, and J. C. Morris. The Policy Implications of Rising Republicanism: Social Welfare Reform in the Deep South. Paper presented at the 1998 Citadel Symposium on Southern Politics, March 5–6, Charleston, South Carolina.

———. 2002. Welfare Reform—Mississippi Style: TANF and the Retreat from Accountability. *Public Administration Review* 62, 1:92–103.

Breaux, D. A., J. C. Morris, and R. Travis. 2007. Explaining Welfare Sanctions and Benefits in the South: A Regional Analysis. *American Review of Politics* 28, 1:1–18.

Breen, J., and P. Anastas. 1994. DfE: The Environmental Paradigm for the 21st Century. *Proceedings of the Symposium of American Chemical Society Committee for Environmental Improvement, August 21–25, 1994*. Washington, DC: US Environmental Protection, Office of Pollution Prevention and Toxics.

Bryner, G. C. 1995. *Blue Skies, Green Politics*. Washington, DC: CQ Press.

Carson, R. 1962. *Silent Spring*. Boston: Houghton Mifflin.

Dietrich, W. 1995. The Nature of Our Future. *Seattle Post-Intelligencer*, March 26, 1995.

Dower, R. C. 1990. Hazardous Wastes. In *Public Policies for Environmental Protection*, edited by P. R. Portney. Washington, DC: Resources for the Future.

Dror, Y. 1968. *Policymaking Reexamined*. Scranton, PA: Chandler.

Elazar, D. 1984. *American Federalism: A View from the States*, 3rd ed. New York: Harper Row.

Fiorino, D. J. 2006. *The New Environmental Regulation*. Cambridge, MA: MIT Press.

Freeman, A. Myrick III. 1990. Water Pollution Policy. In *Public Policies for Environmental Protection*, edited by P. R. Portney. Washington, DC: Resources for the Future.

Gunningham, N., and P. Grabosky. 1998. *Smart Regulation: Designing Environmental Policy*. Oxford: Oxford University Press.

Heilman, J. G., and G. W. Johnson. 1992. *The Politics and Economics of Privatization: The Case of Wastewater Treatment*. Tuscaloosa, AL: University of Alabama Press.

Kaufman, S. 1995. *At Home in the Universe*. New York: Oxford University Press.

Key, V. O. 1949. *Southern Politics in State and Nation*. New York: Knopf.

Kraft, M. E. 2006. Environmental Policy in Congress. In *Environmental Policy: New Directions for the 21st Century*, edited by N. J. Vig and M. E. Kraft. Washington, DC: CQ Press.

Lentz, J. M., and P. Leyden. 1996. RECLAIM: Los Angeles' New Market Based Smog Cleanup Program. *Journal of the Air and Waste Management Association* 46:195–206.

Morris, J. C. 1996. Institutional Arrangements in an Age of New Federalism: Public and Private Management in the State Revolving Fund Program. *Public Works Management & Policy* 1, 2: 145–57.

———. 1997. The Distributional Impacts of Privatization in National Water Quality Policy. *Journal of Politics* 59, 1: 56–72.

Morris, T. 1994. *True Success*. New York: Putnam.

Portney, P. R. 1990. Air Pollution Policy. In *Public Policies for Environmental Protection*, edited by P. R. Portney. Washington, DC: Resources for the Future.

Quarles, J. 1995. *American Environmental Regulation: Brief for Reform*. Washington, DC: Pfizer Inc.

Rabe, B. 2006. Power to the States: The Promise and Pitfalls of Decentralization. In *Environmental Policy: New Directions for the 21st Century*, edited by N. J. Vig and M. E. Kraft. Washington, DC: CQ Press.

Ringquist, E. J. 1993. *Environmental Protection at the State Level: Politics and Progress in Controlling Pollution.* Armonk, NY: ME Sharpe.

Rosenbaum, W. A. 2005. *Environmental Politics and Policy.* Washington, DC: CQ Press.

Scheberle, D. 1997. *Federalism and Environmental Policy: Trust and the Politics of Implementation.* Washington, DC: Georgetown University Press.

Schmidheiny, S., and F. Zorraquin. 1996. *Financing Change: The Financial Community, Eco-Efficiency and Sustainable Development.* Cambridge, MA: MIT Press.

Shapiro, M. 1990. Toxic Substances Policy. In *Public Policies for Environmental Protection*, edited by P. R. Portney. Washington, DC: Resources for the Future.

Spence, D. B. 1995. Paradox Lost: Logic, Morality, and the Foundations of Environmental Law in the 21st Century. *Columbia Journal of Environmental Law* 20, 1:145–82.

U.S. Council on Environmental Quality. 1993. Environmental Quality—1992. Washington, DC: U.S. Government Printing Office.

U.S. EPA Office of Air Quality Planning and Standards. 2005. National Air Quality and Emission Trends Report. Washington, DC: U.S. Government Printing Office.

U.S. EPA Office of Water. 1992. National Water Quality Inventory Report to Congress. Washington, DC: U.S. Government Printing Office.

Von Bismark, O. *Remark August 11, 1867.* http://quotationspage.com/quotes/Otto_Von_Bismarck/ (accessed on February 14, 2007).

Wiener, J. B. 1995. Law and the New Ecology: Evolution, Categories and Consequences. *Ecology Law Quarterly* 22, 2:325–57.

2

State Commitment to Environmental Quality in the South

A Regional Analysis

David A. Breaux, Gerald Andrews Emison, John C. Morris, and Rick Travis

INTRODUCTION

State governments matter substantially to the success of national environmental protection efforts. What the states do, how they do it, and the effectiveness they achieve in environmental protection, are matters of substantial consequence to public policy in the United States (NAPA 1997). As a result, the commitment states have to environmental protection is of surpassing importance. Understanding the sources of such commitment and the influences that moderate this commitment is vital. It can yield insights into advancing environmental protection. Further, it can deepen our understanding of states' general policy implementation effectiveness and thereby improve our knowledge of a pivotal set of actors in the general national policy process. Expansion of state responsibilities brings with it an expansion in the demands for state resources to meet these responsibilities. State governments now conduct most environmental planning, issue regulations, manage individual emission permits, and conduct enforcement activities that pursue the effective implementation of regulatory decisions. Progress in national air quality, water quality, and hazardous waste management now depends on the commitment of state environmental institutions.

This is particularly true in southern states, which have traditionally lagged behind the rest of the nation in environmental protection, both in terms of the implementation of federal standards and in the adoption of state standards. While the nature and severity of environmental degradation may differ in the South, the way southern states respond to federal environmental initiatives is ripe for examination.

19

This chapter examines the broader influences on state environmental capabilities. We develop and test a model that seeks to explain the factors that determine a state's commitment to environmental protection. We also seek to determine whether the factors that explain a fifty-state model are equally important in southern states. If V. O. Key (1949) was correct, we should expect to see differences between southern states and non-southern states. On the other hand, if Steed, Moreland, and Baker (1990) are correct that the South is becoming somewhat less distinct, then we should see relatively minor differences between southern states and the rest of the nation. Understanding the influences on this capacity can yield specific policy insight into sustaining environmental performance in the United States. Further, it can advance our understanding of the roles that various factors have upon more general state capabilities. This is necessary in order to understand capacity influences, to avoid false starts in funding further environmental activities, and to be efficient with new policy efforts at the state level.

INTERGOVERNMENTAL NATIONAL ENVIRONMENTAL POLICY: A PRIMER

Environmental protection in the United States is an intergovernmental venture. It presents a complicated mixture of national, state, and local requirements. Each level of government plays a role in establishing regulatory standards, planning for implementation of the standards, specifying how environmental quality will be measured and monitored, and taking enforcement action when monitoring reveals regulatory standards have been exceeded. The sharing of these responsibilities, however, is not static (NEPI 1997).

Environmental regulation in the United States today is shifting emphasis from the national level to the state level. Such a shift heightens the importance of the performance of state environmental organizations. Environmental progress depends upon state governments' commitment to identify, plan for, and implement environmental regulatory programs. This shift in responsibilities makes state environmental commitment a central feature of successful environmental progress in the United States. Understanding the sources and influences on state environmental capacities can shed light on the performance improvements at the state level that will be necessary as these responsibilities devolve more and more from the federal level to the state level.

States' Roles as Envisioned by National Environmental Statutes

The Clean Water Act, the Clean Air Act, and the Resource Conservation and Recovery Act contain similar expectations for the division of responsibilities between the federal and state governments (Rosenbaum 2006).

Legislative expectations for these statutes were straightforward. The federal government would identify national standards, and the states would see that those standards were implemented. If the state implementation was inadequate, the federal government was authorized to act instead. Statutory history reveals an intended approach of centralized standards setting and decentralized implementation for the nation's environmental regulatory programs (NAPA 1995). These statutes contemplated that the states would see that federal standards were achieved through state level planning, permit issuance, environmental monitoring, and conduct of compliance enforcement operations. The statutes were generally silent about the relationship between the state and local governments, leaving that to the discretion of state legislatures and administrators. Primary attention was focused on the division of responsibilities between the federal government and the state governments. It is especially noteworthy that in the event a state program was inadequate, either from a design or an implementation standpoint, the federal government was statutorily directed to undertake program implementation in place of the state. But this was clearly not the preferred role for the federal government.

States' Roles as Implemented

The expectations for robust state implementation of national standards were too optimistic. They did not take into account the substantial fiscal, managerial, and political capacity building that would be necessary before the state governments could assume the role the national statutes envisioned. For most programs sponsored by national environmental regulatory legislation, the devolution of environmental regulation to the state level has taken longer than expected (NAPA 1997). State governments lacked the necessary statutory authority, the institutional structures, the managerial expertise, and the technical capabilities to carry out many of the operations the statutes envisioned. This prompted EPA, acting under the default provisions of the statutes, to assume substantial implementation responsibilities. A prolonged period of overlapping responsibilities ensued. In many instances, such as often was the case with the national pollutant discharge elimination system of the Clean Water Act, EPA retained operational decision-making authority and the state governments often functioned as first level staff analysts and engineers. In other cases, such as hazardous waste disposal permits, EPA retained the entire permitting process and state governments served as commentators on the EPA-proposed permitting decisions.

Sources of Change in Intergovernmental Responsibilities

A combination of factors is acting to change the intergovernmental relationships that originally characterized environmental regulatory decisions. The results are relationships that align more consistently with the original statutory intent. State political leaders have learned of the enormous economic, political, social, and environmental importance of operational environmental decisions. They generally believe that such decisions are best made by state rather than federal officials. At the same time, state environmental agencies have grown in managerial and technical expertise. When these capabilities are paired with the institutional capabilities associated with strengthened state environmental statutes, a more muscular state environmental capability emerges. The shift to a more robust state environmental interest has been prompted also by necessity. The initial regulatory challenge was to control large, discrete point sources of pollution. These are sources that can be identified and controlled through direct federal regulation. Today, these large sources are generally regulated, and smaller, more dispersed and nonpoint sources constitute the emissions inventory to be controlled. Such sources do not respond well to federal regulation, but are more amenable to control through the more hands-on, direct approaches available only through state governments. This trend has been reinforced by limitations imposed by federal budgetary constraints and the political view at the national level of limiting the role of the federal government.

Increasing state capacity, changes in the character of emission sources, a shift in the philosophy of federal involvement, and limitations on the growth of the federal budget, accompanied by more state legislative desire for control over environmental regulation, have yielded a climate in which the expansion of state participation in environmental regulation is persistent and accelerating. In air pollution control the state implementation plan is the principal mechanism for designing and managing ozone and particulate matter reductions, and these plans are developed and implemented at the state level with EPA approval. In the water quality program the development of total maximum daily loads for impaired water bodies has become the principal mechanism for assigning control requirements in critical watersheds. The state governments are the principal organizations responsible for developing Total Maximum Daily Loads, and EPA has sought to remove itself from this process (Rosenbaum 2006).

Our goal is thus to determine the factors that explain the choices states make in terms of environmental protection. We develop a model that tests some of the more common explanators found in the state comparative literature. Furthermore, we seek to determine whether the importance of these explanators differs between southern states and the rest of the nation. The following section specifies the variables employed in our models.

VARIABLE SPECIFICATION

James Lester (1994) provides a useful summation of much of the work done to explain state variation in the implementation of environmental policy. The severity argument is predicated on the differences in the degree of environmental problems within states. States with more pressing environmental problems will move quickly, while states with less severe problems lack the motivation to move quickly. However, Lester also states that previous research indicates that "the correlation of severity to protection is not clear, and that more refined indicators of pollution severity . . . are needed . . . [f]actors other than problem severity (such as politics or economics) appear to be affecting states' behavior in this area" (59). On the other hand, Lester argues that the wealth hypothesis can explain a significant amount of the variation in state protection efforts. The partisanship argument, according to Lester, is the most common explanation found in the literature (60)—that ideology or political party identification can explain implementation variation. Calvert (1989) has found some evidence for this explanation. The culture argument, as laid out in chapter 1, contends that southern states have unique cultural norms that impact policy decisions. Finally, the organizational capacity argument suggests that factors such as gubernatorial control, professional legislatures, and consolidated environmental bureaucracies (Lester 1994, 62) can be used to explain state variation in implementation, and that there is some evidence to support this view.

The relationship between state environmental spending and environmental quality is ambiguous. A number of authors have advanced the concept that environmental quality is influenced by environmental protection activities. These activities are in turn influenced by state and federal expenditure levels (Rosenbaum 2006; Fiorino 2006). Such a logical path of influence was challenged by Ringquist (1993). His research questioned the causal role of expenditures at the state level upon environmental results. In this research on air and water quality outcomes, a wide range of variables such as regulatory program strength, industrial output, and regulatory prescriptions appear to exert influence rather than state expenditures. Nevertheless, the role of resources in conditioning environmental performance continues to exert a powerful draw to explain state environmental capacity and results.

In addition to pollution severity, state policy choices are also presumed to be impacted by political considerations. Pluralism theory offers the argument that interest groups can play an important role in influencing policy outcomes dependent on their strength. In the area of environmental management a variety of scholars have focused on this issue (Bacot and Dawes 1997; Newmark and Witko 2007; Ringquist 1993) and have found mixed evidence concerning the influence of environmental and industrial interest groups. For example, Ringquist (1993) found that the influence of these interest groups varied depending on the specific environmental policy area. These groups mattered greatly concerning water policy adoption but have little role to play with air quality issues. Newmark and Witko (2007) found that environmental groups had more influence in policy making than did industrial groups. Specifically, they found that a positive relationship exists between the size of environmental group membership in a state, as measured by membership in the Sierra Club, and the level of spending on environmental and natural resource protection. This result is consistent with Bacot and Dawes (1997), who found that environmental interest groups significantly impact state environmental expenditures while industrial interest groups had no impact. For an index measuring a broader definition of environmental effort, they found that neither type of interest group had a significant impact. In this chapter we will use the data taken from Bacot and Dawes who measure the strength of interest groups at the state level in 1987. Their measures relied on state level membership in the Sierra Club on a per capita basis and per capita membership in the American Chemical Society, the National Association of Manufacturers, and the Society for Mining, Metallurgy, and Exploration to operationalize environmental and industrial group strength respectively (1997, 359–60).

Other political factors are also hypothesized to be related to policy outcomes (Feigenbaum and Henig 1997). The conventional argument is that Democratic controlled states are more likely to be supportive of environmental efforts while Republican controlled states are less likely. At a regional level, however, we are less certain that such an argument would hold for democratically controlled states in the South. Owing to its traditionally conservative roots, the Democratic Party in the South is less apt to be supportive of environmental efforts when compared to the Democrats nationally. Thus, we expect that as the dominance of the Democratic Party over a state outside the South increases, the more likely a state is to support environmental efforts. We use the interparty competition index, developed by Austin Ranney and modified by John Bibby and others (Bibby et al. 1990), to measure party dominance in the states. Measured for the 1981–1988 period, this index considers the success of each party in winning control of state legislative and gubernatorial seats as well as the number of votes that each party typically won in such elections. A score of 1 indicates complete Democratic dominance while a score of 0 indicates complete Republican control.

In addition to these partisan measures, we add a measure of political ideology. We selected the citizen ideology scores drawn from Berry et al. (1998) for 1987 where higher scores indicate a more liberal ideology. This variable helps us to capture a broader component of citizen political preferences than just focusing on party control. As Berry et al. note, by including the preferences of citizens who vote for losing candidates in their measure they are better able to capture more nuances in ideology. We expect that states that are more liberal in ideological orientation are more likely to adopt supportive environmental policies.

Another set of factors purported to be related to environmental effort is the economic health of a state. For many states, historically, resource and environmental protection has been a secondary or even tertiary policy concern. Many states have typically viewed their most important priorities as consisting of education, economic development, and the maintenance of law and order. Research in the environmental arena has shown a relatively consistent relationship between state economic health and environmental support (Bacot and Dawes 1997; Goggin et al. 1990).

To capture the economic well-being of the states we use two measures. The first focuses specifically on the fiscal health of the state government, the second on the wealth of a state. To measure the fiscal health of a state, we use Tannenwald's fiscal comfort index for 1987. The fiscal comfort index is an interval level measure comparing the index of tax capacity of a state to the index of fiscal need of the state (Tannenwald 1999). By using this measure, we gain a picture of the fiscal environment state governments were facing as they made environmental policy decisions. The second variable measures gross state product per capita. Taken from the *State Politics and Policy Quarterly* (2007) states dataset, this variable goes beyond measuring just the position the state government finds itself in fiscally and investigates the idea that environmental protection, as a tertiary concern, is something that can be "afforded" in richer states. In sum, we expect that both variables are positively related to environmental efforts.

In addition to the above factors we also investigate the possibility that environmental efforts are tied to the administrative abilities of states. Lester (1994) argues state policy activity is influenced by a rich intergovernmental framework that includes these state factors, as well as federal-level inducements and constraints (see also Goggin et al. 1990). Given the movement toward devolution and decentralization of federal environmental policy, he argues that a state's willingness to implement federal policy is fundamentally the result of two factors: the state's overall commitment to the environment, and the institutional capacity of the state to implement complex federal programs (Lester 1994, 62–63).

Institutional capacity, or state capability, is measured using interval level data from Bowman and Kearney (1988), as suggested by Lester (1994). Bowman and Kearney calculated factor scores for four dimensions of state capacity for each state (staffing and spending, accountability and information management, executive centralization, and representation). For our study, we will use the summed scores of these four dimensions as our measure of government capacity. As Lester notes, those states with a greater capacity are expected to be more assertive in environmental protection

The structure of the state-level agency designed to coordinate environmental policy provides another organizational measure of a state's commitment to the environment. Using the Bacot and Dawes measurement of state environmental organizational structures, the dichotomous level measurement categorizes state environmental agencies into an EPA-like or superagency structure (Bacot and Dawes 1997). The EPA-like structure follows the EPA organizational structure, while the superagency is a larger, more diverse environmental organization. Research suggests that the EPA structure is better able to address federal environmental program responsibilities and to ensure intergovernmental consistency than the superagency structure (Goggin et al. 1990). Because the adoption of an EPA-like structure can be viewed as an evolutionary step forward bureaucratically we would expect states with an EPA-like structure to be more supportive of the environment.

Finally, the role of political culture and the argument that the South is unique are also tested. As Key (1949) has noted the South historically has been dominated by a conservative political culture that has led it to be seen as "unique" concerning some of its policy choices. Elazar defined the eleven states of the Old Confederacy plus West Virginia, Kentucky, Oklahoma, Arizona, and New Mexico as possessing a traditional culture that lent itself to the maintenance of a hierarchical political, social, and economic system and the preservation of the status quo to the detriment to a willingness to adopt new policy positions. In order to detect regional differences that may be related to culture we add six additional independent variables to those discussed above that allow us to measure the distinctive impact that we expect for the southern states.[1] Specifically, we test the hypotheses that the effects of pollution, interest groups, party competition, and fiscal and financial variables have differential impacts for the non-southern states compared to the southern states. By calculating the interaction of these six variables with a dummy variable measuring the South and adding only the interactive terms to the original variables we are able to sort out distinctions between the South and non-southern concerning the determinants of environmental commitment.[2]

METHODS

Each of the hypotheses developed in the preceding section is operationalized and tested in the models presented in the following section. The two dependent variables include a measure of financial commitment on the part of a state to addressing its environmental needs and a second measure which is a more holistic approach capturing financial, political, and administrative support that a state devotes to environmental protection. The first measure operationalizes commitment as a measure of state expenditures on environmental programs on a per capita basis.[3] Financial expenditures are often viewed as the most transparent measure of a state's policy commitments. Absent adequate funding, the adoption of other state regulations and programs can be viewed as relatively hollow efforts (Bacot and Dawes 1997). This operationalization on a per capita basis also has the strength of more directly measuring a state's commitment relative to its size. Financial expenditures, however, are not the only recourse that policy makers have for demonstrating environmental commitment.

The second dependent variable is taken from Davis and Lester (1989) who move beyond simply measuring financial support and also include measures of a state's political and administrative support. For this measure they note that political support "refers to the enactments of several environmental protection policies" and administrative support "refers to the type of institutional structure and personnel available to administer federal environmental programs" (71–72). With this measure we approach environment commitment in a different context. Although Bacot and Dawes' critique about the futility of legislation without funding has a certain level of truth to it, it is not always the case that policy commitments require funding. For instance many environmental and natural resource regulations require relatively low levels of state spending. Further, it is also the case that environmental spending as a measure can be seen as too dependent on the financial health of a state.[4]

RESULTS

Our first model employs the environmental expenditures per capita as the dependent variable (see table 2.1). In this model we analyze forty-nine states using our nine independent variables and the additional six interactive variables designed to capture distinct differences for the South.[5] The results for the full model and also for the reduced model in table 2.1 are relatively disappointing in terms of their explanatory capability with the relatively modest R^2s of .211 and .286 respectively. Additionally, we find that none of the distinct South variables are significant. The three factors that are consistently significant are the financial variables and an institutional variable. As expected, we find a significant positive relationship between fiscal comfort levels and a state's spending on the environment. Consistent in both models, the result indicates that those states that have a better ratio of their tax capacity to their fiscal needs are more apt to fund environmental protection efforts at higher levels. Conversely, those states that are facing a challenging fiscal situation are less inclined to fund environmental programs. What these results cannot answer, however, is the order of tradeoffs that fiscally strapped states make. In other words, our results only tell us that fiscally strapped states spend less than others. They cannot answer the question of whether these states spend less across the board or if environmental protection efforts are indeed viewed

Table 2.1. Environmental Commitment Based on Expenditures Per Capita: 1987

	Full Model		Reduced Model	
	Coefficient	Std. error	Coefficient	Std. error
Pollution Level	28.51	68.98		
South * Pollution Level	−88.82	145.02		
Industrial Group Membership	−.772	5.83		
South * Industrial Group Membership	72.08	74.09		
Environmental Group Membership	1.34	1.66		
South * Environmental Group Membership	.833	20.12		
Citizen ideology	−.009	.313		
Interparty competition	−.194	.253		
South * Interparty competition	1.25	1.59		
Fiscal Comfort	.727***	.244	.769***	.195
South * Fiscal Comfort	−.447	2.40	−.170	.876
Income per capita	−.009***	.002	−.010***	.002
South * Income per capita	.002	.011	.004	.008
Institutional Capacity	4.27*	2.32	4.07**	1.92
EPA-like agency	−2.844	7.282		
intercept	30.07	4.70	28.175	2.96
	N= 49 Adj. R² = .211		N=49 Adj. R² = .286	

*p<.10; **p<.05; ***p<.01

as being of secondary or tertiary importance and therefore are among the first programs to be underfunded.

The observed relationship between income per capita and state environmental expenditures is somewhat counterintuitive. It is consistently negative, which on the surface tells us that richer states spend less per capita on the environment than poorer states when we control for the other variables. In a simple bivariate analysis there is no correlation between income and environmental spending and there is a relatively low correlation between fiscal comfort and income.

Finally, our results indicate support for Lester's argument concerning the relationship between governmental capacity and the willingness to commit resources to environmental protection. In sum, however, our analysis of this dependent variable indicates a relatively clear linkage between a dependent variable focused specifically on fiscal effort and independent variables focused on fiscal abilities. This, perhaps, speaks to the relative narrowness of expenditure per capita as a measure of environmental effort.

In the second model displayed in table 2.2 we find the results of our analysis of the Davis and Lester index of environmental commitment.[6] This analysis does generate some very important regional differences concern-

Table 2.2.   Environmental Commitment Based on Davis and Lester Index

	Full Model		Reduced Model	
	Coefficient	Std. error	Coefficient	Std. error
Pollution Level	25.58	23.73	40.751*	20.482
South * Pollution Level	−98.31**	50.15	−85.430**	34.503
Industrial Group Membership	.209	1.99	−.503	1.802
South * Industrial Group Membership	−44.79	60.11	−59.911*	34.07
Environmental Group Membership	.300	.539		
South * Environmental Group Membership	−5.64	7.92		
Citizen ideology	.279**	.106	.288***	.080
Interparty competition	.019	.087		
South * Interparty competition	.125	.650		
Fiscal Comfort	−.098	.083	−.084	.078
South * Fiscal Comfort	1.73**	.891	1.475*	.735
Income per capita	.001	.001	.002**	.001
South * Income per capita	−.006	.004	−.006	.004
Institutional Capacity	.633	.795		
EPA-like agency	3.16	2.38		
intercept	27.52	1.62	29.30	1.12
	N= 49 Adj. R^2 = .391		N=49 Adj. R^2 = .433	

*p<.10; **p<.05; ***p<.01

ing the impact of pollution severity, interest group membership, and fiscal comfort. Regarding the impact of yearly pollution levels and the Davis and Lester index, the results in the reduced model and, although not significant, the coefficient in the full model indicate a positive relationship between pollution levels and the index of environmental commitment. It seems that the higher the levels of pollutants the more states outside of the South are likely to increase their commitment to natural resource and environmental protection. For the South, however, we find a negative relationship, a finding that is somewhat puzzling to us. We can think of a number of possible explanations for this finding. One group of explanations employs the argument of southern distinctiveness; another view holds the opposite view, that the South is losing its distinctiveness.

First, it could be that in the South the causal relationship is backwards. In other words, whereas in the rest of the country increasing pollution has led states to adopt higher standards, in the South lower environmental standards have allowed pollution to increase. Another way of thinking about this is that the South has both lower levels of economic development and lower levels of pollution when compared to the rest of the country. So, in the South pollution is seen as a side effect of a happy development, of economic growth. In much of the developing world, increasing pollution is seen as a mark of economic maturation. It may thus be that the South is still in more of a materialist mindset where economic concerns are paramount, whereas the rest of the country is more apt to have a postmaterialist mentality where other concerns are more important. In such a circumstance the South's distinctiveness might rest on the pro–small government, pro-business views that characterize an industrial society.

The opposite explanation views the South as losing its uniqueness as one state after another becomes postmaterialist. In this view states of the new South, such as Florida, North Carolina, and Virginia, are becoming more like the rest of the nation as their economies and cultures modernize, and the remainder of the South will present such altered views over time as they move from industrial to postindustrial status. Under such conditions explanations based on southern distinctiveness diminish as southern states join the rest of the nation in postmaterialist traits, one of which is commitment to pollution control. Evidence for such a situation requires examination of attitudes and conditions over time. The research we present herein is that of a single snapshot. Additional time series analyses should yield insight into the validity of this view.

Our results for the full model also indicate a significant negative relationship between industrial group membership and environmental commitments among southern states. Whereas we find no relationship for the rest of the country, in the South we find that as the size of industrial group membership grows, so too, at least potentially, does their lobbying influence to keep environmental regulations low. Still, this finding is only weakly significant in the reduced model.

Finally, our results also indicate a difference between southern and non-southern states concerning the impact of state fiscal comfort on environment commitment. For the non-South we find that no statistical relationship exists between these two variables but for southern states increasing levels of fiscal comfort are positively associated with increasing environmental commitments. In the South is seems that the willingness to adopt more stringent legislation and to invest in enhancing their institutional structure and personnel is related to their ability to afford such steps. After all, the two southern states with the lowest environmental scores, Alabama and Mississippi, are also the two states with the worst fiscal comfort levels.

In addition to these three variables where we find significant differences in southern and non-southern states, we also find that significant, positive relationships exist between the measures of citizen ideology and income per capita and environmental commitment. Both of these findings are in the expected direction with states that have a more liberal ideological score and that are richer being more apt to demonstrate a higher commitment to the environment. For both of these variables the relationship among southern states is consistent with the rest of the nation.

CONCLUSIONS

Our findings indicate four relatively important conclusions. First, our analyses suggest that the South is indeed different from the non-South in some important ways. Perhaps the most striking of these differences is indicated by the significance of pollution levels, industrial group membership, and fiscal comfort. While fiscal comfort and environmental commitment are negative and insignificant for all states, it appears that higher levels of fiscal comfort in southern states do lead to a higher level of environmental commitment (as measured by the Davis and Lester index). Likewise, it appears that the lack of a strong environmental lobby in the South provides additional influence for industrial groups, who are likely to oppose environmental protection efforts.

An area that needs further theoretical and empirical development is the observed relationship between environmental commitment and pollution levels. While the relationship is positive (and weakly significant) for all states, the finding of a strong inverse relationship for southern states seems counterintuitive. As noted earlier in this chapter, we can think of a couple of potential explanations for the observed relationship, but further research is needed to better develop the theory behind this finding. We wonder whether it is the case that the South is losing its distinctiveness over time (see Steed, Moreland, and Baker 1990), or whether the development mindset in the South is simply stronger than in other regions of the nation.

Second, in other important ways the South is not different than the non-South. Our results indicate that political factors, namely the influence of environmental interests groups and interparty competition, are not related to environmental commitment among either southern or non-southern states. This begs the question about the applicability of the "politics matters" thesis. In our analysis, at a region-wide general environmental commitment level, politics does not matter much. Perhaps though, it is the case that the role of politics is better illustrated by focusing on more specific environmental issues and on more specific policy decisions. This is one of the themes for other chapters to take up later.

Third, measuring environmental commitment by per capita expenditures, as suggested by Bacot and Dawes (1997), is likely plagued by at least two issues. As noted earlier, such a measure cannot capture the tradeoffs states make in how to spend their scarce resources. In addition, unless longitudinal data are employed in the analysis, there is no way to determine the degree to which a single datum adequately captures a state's fiscal commitment. We thus conclude that measuring a state's environmental commitment in this manner is, at best, of limited utility.

Finally, there is clearly additional work needed to better capture the differences in state commitment to the environment. Our best model explains just over 40 percent of the variance; we suspect that by incorporating some of the nonenvironmental independent variables that have shown some explanatory power in other policy arenas (see, for example, Breaux, Morris, and Travis 2007; see also Soss et al. 2001), we might be able to increase our ability to explain variance in the dependent variable, as well as detect additional differences between southern states and the rest of the nation. This work is the subject of future research.

NOTES

1. Our "South" definition includes the states of North Carolina, South Carolina, Georgia, Florida, Alabama, Mississippi, Tennessee, Texas, Louisiana, Arkansas, and Virginia. Thus, it is not directly coterminous with Elazar's identification of traditional states, but this does allow us to test the notion of the South's cultural uniqueness.

2. In the analyses with interactive terms the original variables are centered before creating the interactive terms. This makes for easier interpretations of the results (Friedrich 1982).

3. Expenditure data for 1987 is taken from Bacot and Dawes (1997). In their work they used gross expenditures to measure financial commitment whereas we use a per capita measure. Additionally, the state population data used to per capitalize the measure is taken from the same source. This variable is measured in thousands of dollars per person.

4. Indeed the two dependent variables used in this chapter have a correlation of −.03 indicating that they are not measuring the same thing. Further consideration of how to measure environmental effort seems to be warranted

5. In this model Alaska is identified as an outlier and is dropped from the analysis predominately because of its very high value on the dependent variable. Alaska's pollution expenditure is $323 dollars per capita while the next highest state, Wyoming, is only $134. The mean value is $33.18 per capita.

6. In the models for table 2.2, Florida was detected as an outlier using Cook's D and DFBETAS statistics. Thus it was removed from this analysis. See Fox (1991) for more details. Florida is predominately an outlier based on some of the independent variables, especially its industrial group membership which is unusually high and its fiscal comfort, which is also high for the southern region.

REFERENCES

Bacot, H. A., and R. A. Dawes. 1997. State Expenditures and Policy Outcomes in Environmental Program Management. *Policy Studies Journal* 25, 3:355–70.
Berry, W. D., E. J. Ringquist, R. C. Fording, and R. L. Hanson. 1998. Measuring Citizen and Government Ideology in the American States, 1960–93. *American Journal of Political Science* 42, 1:327–48.

Bibby, J. F., C. Cotter, J. Gibson, and R. Huckshorn. 1990. Parties in State Politics. In *Politics in the American states*, 5th ed., edited by V. Gray, H. Jacob, and R. Albritton, 66–113. Glenview, Ill: Scott Foresman.

Bowman, A. O'M., and R. C. Kearney. 1988. Dimensions of State Government Capability. *Western Political Quarterly* 41:341–62.

Breaux, D. A., J. C. Morris, and R. Travis. 2007. Explaining Welfare Benefits in the South: A Regional Analysis. *American Review of Politics* 28, 1:1–18.

Calvert, J. W. 1989. Party Politics and Environmental Policy. In *Environmental Politics and Policy: Theories and Evidence*, edited by J. P. Lester, 158–78. Durham, NC: Duke University Press.

Davis, C. E., and J. P. Lester. 1989. Federalism and Environmental Policy. In *Environmental Politics and Policy: Theories and Evidence*, edited by J. P. Lester, 57–84. Durham, NC: Duke University Press.

Feigenbaum, H. B., and J. R. Henig. 1997. Privatization and Political Theory. *Journal of International Affairs* 50, 2:338–55.

Fiorino, D. J. 2006. *The New Environmental Regulation.* Washington, DC: CQ Press.

Fox, J. 1991. *Regression Diagnostics: An Introduction* (Sage University Paper Series on Quantitative Analysis in the Social Sciences, 07-079). Newbury Park, CA: Sage.

Friedrich, R. J. 1982. In Defense of Multiplicative Terms in Multiple Regression Equations. *American Journal of Political Science* 26:797–833.

Goggin, M. L., A. O'M. Bowman, J. Lester, and L. J. O'Toole, Jr. 1990. *Implementation Theory and Practice: Toward a Third Generation.* Glenview, Ill: Scott, Foresman/Little.

Key, V. O. 1949. *Southern Politics in State and Nation.* New York: Knopf.

Lester, J. P. 1994. A New Federalism? Environmental Policy in the States. In *Environmental Policy in the 1990s*, 2nd ed, edited by N. Vig and M. E. Kraft, 51–69. Washington, DC: CQ Press.

National Academy of Public Administration (NAPA). 1995. *Setting Priorities, Getting Results: A New Direction for EPA.* Washington, DC: Author.

———. 1997. *Resolving the Paradox of Environmental Protection: An Agenda for Congress, EPA and the States.* Washington, DC: Author.

National Environmental Policy Institute (NEPI). 1997. *Environmental Goals and Priorities: Four Building Blocks for Change.* Washington, DC: Author.

Newmark, A. J., and C. Witko. 2007. Pollution, Politics, and Preferences for Environmental Spending in the States. *Review of Policy Research* 24, 4:291–308.

Ringquist, E. J. 1993. *Environmental Protection at the State Level.* London: Sage.

Rosenbaum, W. 2006. *Environmental Politics and Policy*, 7th ed. Washington, DC: CQ Press.

Soss, J., S. S. Schram, T. P. Vartanian, and E. O'Brien. 2001. Setting the Terms of Relief: Explaining State Policy Choices in the Devolution Revolution. *American Journal of Political Science* 45, 2:378–95.

State Politics and Policy Quarterly. 2007. State Politics and Policy Quarterly Data Resource. www.ipsr.ku.edu/SPPQ/datasets.shtml (accessed February 19, 2008).

Steed, R. P., L. W. Moreland, and T. A. Baker (eds.). 1990. *The Disappearing South? Studies in Regional Change and Continuity.* Tuscaloosa, AL: University of Alabama Press.

Tannenwald, R. 1999. Fiscal Disparity among the States Revisited. *New England Economic Review* (July/August 1999):3–24.

3

Ozone Air Quality Management and Southeastern Distinctiveness

Gerald Andrews Emison

INTRODUCTION

Controlling ground level ozone air pollution is a major environmental concern in the United States. This air pollutant causes or contributes to substantial human health impacts as well as a wide range of adverse environmental effects. Because ozone pollution results from photochemical reactions involving compounds extensively used in modern industrial society, Americans, particularly those in urban areas, are exposed to high levels of ozone. The ubiquity and adverse health effects of ozone make control of this pollutant both imperative and expensive. This chapter examines the control of ground level ozone among the southeastern states and measures this region's results against those from the rest of the United States. In doing so it identifies important differences that exist between the Southeast and the rest of the United States. Examination of these differences suggests that they exist in part due to southeastern states' lower than national funding for environmental protection rather than for specific air quality programs.

BACKGROUND

Ozone Air Quality as a Major Public Issue

The human health effects of ozone pollution are well documented. Ozone oxidizes lung passages, damaging alveoli through burning the tissue. While anyone may be adversely affected by exposure to ozone, individuals with respiratory problems and healthy individuals exercising heavily are most likely to have their lungs damaged as a result of ozone exposure. Permanent lung damage can result from prolonged exposure to ozone, exacerbating asthma, diminishing lung capacity and heightening susceptibility to bronchitis and pneumonia. In addition to human health effects, it is well established that ozone can interfere with plants' abilities to produce and store food. This diminishes the survivability of vegetation subjected to extreme weather, plant disease, and pests (U.S. Environmental Protection Agency 2004). As a result ozone has been identified as a contributor to the reduction of a large number of leafy agricultural crops as well as to adverse affects on forests.

The precursors to ozone, volatile organic compounds and oxides of nitrogen, are typically emitted en masse by human activities concentrated in our cities. Such conditions have resulted in widespread nonattainment for our urban areas of the health-based national ambient air quality standard for ozone. The U.S. Environmental Protection Agency estimates that well over 100 million Americans are exposed to air contaminated with ozone at an unhealthy level (U.S. Environmental Protection Agency Office of Air Quality Planning and Standards 2004a).

Initially ground level ozone was thought to be a major environmental concern only in urban areas. Further scientific studies have shown that ozone concentrations in nonurban areas approach, and in some cases, exceed levels of concern as a result of ozone transport from urban areas (U.S. Environmental Protection Agency Office of Air Quality Planning and Standards 1996, 1997). As a consequence, ozone control has moved from a concern for urban areas to a concern across the entirety of a state.

The combination of substantial ozone health effects and extensive exposure to such health effects has made ozone control a major focus of state and federal air pollution efforts. A complex and extensive set of governmental regulations has grown up around ozone control (Revez 1997). Emissions standards for particular industries and motor vehicles associated with a generation of ozone precursors as well as ambient standards for ozone concentration receive a substantial portion of the U.S. Environmental Protection Agency's air quality control program attention. Policies to implement these standards are executed through EPA's ten regional offices, providing a degree of regional tailoring. Adoption and implementation of control programs through source inspections and enforcement are the purview of state environmental agencies. Regulations on motor vehicles and industrial sources alike have grown progressively more restrictive in order to deal with ozone air pollution. Such restrictions, not surprisingly, have substantial economic implications by imposing additional costs for emission controls on sources of the precursors of ozone (Rosenbaum 2008).

The consequence of such regulatory activity has been to make ozone control a potential cost to industries and citizens in areas of high ozone pollution. As a result of this confluence of factors, ozone control programs' design and implementation are of substantial interest and concern to local and state economic as well as environmental interests. Any region of the country with major intentions to expand industrial activity will find itself required to address ozone pollution.

In the southeastern United States, in particular, controlling smog is a significant air pollution challenge. The rapid expansion of urban areas along with traditional reluctance to advance restrictive government regulatory programs results in heightened demand for air pollution control that is countered with cautious regulation.

This chapter examines the status of ozone pollution control in the United States in order to learn whether the Southeast's ozone control efforts are unique or align with those of other states. In doing so it examines the role that regional differences play in effectiveness of ozone control as well as the influences upon effectiveness of ozone air quality management at the state level. The public health importance of ozone exposure, the increasing importance of state air quality management decisions, and the latitude of decisions at the subnational and regional levels of EPA indicate that future effective air quality management depends upon understanding regional differences in the relationships between outcomes and influences.

COMPLEX INFLUENCES ON
STATE OZONE CONTROL MANAGEMENT

The literature of environmental management suggests that there are potentially a large number of factors that might influence a state's ozone control effectiveness (Ringquist 1993). There are a large number of actors in the regulatory program, and each actor's influence is expressed through a number of potential pathways. As a consequence, a state government's influence on ozone levels may depend not only upon the state government, but on the motivation to comply with regulatory action on the part of industrial sources, interest groups, and political actors. In short, both capacity and setting may matter, so when we examine state environmental activities we must consider both aspects.

Reflecting this situation, previous research in state environmental effectiveness (Ringquist 1993, Lowry 1992) suggests that a state's action on environmental matters may be influenced, individually or in combination, by its politics, the alignment of its policies, and the nature of the state's economy, while the level of state expenditures on specific environmental programs does not appear to exert appreciable effects on outcomes.

RESEARCH METHODS AND RESULTS

Measuring State Ozone Control Progress

In order to examine progress in ozone control among the states, we need a stable metric by which to measure such progress. State governments are the primary institutions responsible for air quality management planning and implementation (Rosenbaum 2008), so states are the appropriate unit of analysis. They have pursued ozone control by focusing on urban areas. As a consequence EPA and the states collect and report air quality measurements by Metropolitan Statistical Areas (U.S. Environmental Protection Agency Office of Air Quality Planning and Standards 2004a). Metropolitan statistical areas (MSAs) were chosen as the initial unit of analysis for air quality since such areas have individual state implementation plans developed for them for ozone.

Measurement of air quality results were examined using the EPA-developed air quality index (AQI). In brief, the ozone AQI represents the degree to which the monitored air quality in a MSA meets the National Ambient Air Quality Standard (NAAQS) for ozone. An AQI of 100 indicates monitored air quality that is just at the ozone NAAQS of 12 ppm monitored over an eight hour period, the national ambient air quality standard in effect during the period of analysis (U.S. Environmental Protection Agency Office of Air Quality Planning and Standards 2004b). This is a consistent measure of ozone progress across time and geography, hence its stability is attractive from an analytic standpoint.

For the period 1994–2003 each MSA in the nation was examined for its level of the ozone AQI. MSAs were excluded that had less than 350 days of monitored ozone data or did not report data for both 1994 and 2003. The median air quality index during each year was reported for each MSA.

The measure of results performance was developed as the change from 1994 to 2003 in each MSA's median air quality index. In order to gain a picture of the overall performance of a state in its ozone air quality control program the mean performance of all MSAs in a state was computed from the individual MSA data.

Measuring State Influences on Ozone Control

Influences on state ozone control effectiveness were examined for a range of categories. Measures of political alignment were those employed in Erickson, Wright, and McIver's *Statehouse Democracy* (1993). This characterization employed three measures. Strength of partisanship index was derived from CBS/*New York Times* interviews of citizens across the period 1976–1988. The index is the difference of the percentage of respondents identifying themselves as Democrats and those identifying themselves as Republicans. The ideological index was developed in a similar manner: it is the percentage of the respondents in a state identifying themselves in the CBS/*New York Times* polls as liberal minus the percentage identifying themselves as conservative. The policy liberalism index developed by Erickson, Wright, and McIver (1993) is the standardized score of the state's respondents for characterization concerning consumer protection, criminal justice, legalized gambling, the Equal Rights Amendment, and tax progressivity. As a result, the policy liberalism index can be viewed as a general measure of the state's orientation to liberal policies.

Interest group strength was measured by two variables to capture the two competing interests in environmental policy at the state level. Industry interest group strength was measured as the number of manufacturing interest groups registered in each state in 1990 (Lowery 1998). Environmental interest group strength was expressed as the number of Natural Resource Defense Council members in a state in twenty-four months prior to September 1995 (NRDC 1995). Both were adjusted to account for population of a state.

Economic character of the states was captured by measures of aggregate and per capita state economic size. The gross state products for 1996 were obtained from the Bureau of Economic Analysis (2003).

Data concerning state expenditures on overall environmental program expenditures as well as those for air quality were obtained from a survey of state environmental agencies conducted by the Environmental Commissioners of the States (Environmental Commissioners of the States (ECOS) 2004). These represent the expenditures that states reported on their air quality programs in the year 2000, which is approximately the midpoint year for the ozone air quality results data. These were also adjusted for the states' 2000 populations (U.S. Census Bureau 2003).

RESEARCH QUESTIONS

Examining states' effectiveness in controlling ozone involves a series of questions. These questions begin with the descriptive: What is the status of ozone control in the United States? Is there a difference in ozone control among the states? The research next examines the explanatory: What factors might account for the differences in ozone control among the states?

These questions allow deployment of the theories concerning sources of state effectiveness discussed earlier. In particular, they suggest that a series of models built around the principal potential causal influences are appropriate. As a result changes in ozone levels were modeled by employing, separately then in combination, location, economics, policy preferences, political alignment, and interest group strength.

THE CURRENT OZONE SITUATION

Across the period of the analysis the geographic distribution of ozone non-attainment in the United States in 2003 was extensive. The standard employed in determining nonattainment is one of 0.12 ppm ozone averaged over one hour, the standard in effect at the time. The key ozone areas mirror the population concentration, with the Northeast, California, and the Central Midwest presenting the majority of the areas. The Southeast contains a number of such areas, generally those metropolitan areas associated with substantial growth. When trends in ozone control are examined however, a national reduction of 22 percent in the period 1983–2002 can be contrasted with that of between 11 to 17 percent for the Southeast, depending upon which EPA regional office is examined (EPA 2004). The typical concentration of ozone in the Southeast in 2002 was quite similar to that of the United States as a whole. Since most southeastern ozone concentrations in 1983 were lower than non-southeastern areas, this suggests that the Southeast's progress has lagged that of other areas of the country. With less of an ozone problem originally, the Southeast caught up to the rest of the United States through attenuated improvement. When ozone improvement is modeled using geographic location as an explanatory variable (table 3.1), a southern location appears as a significant variable.

Factors Influencing State Ozone Control

These differences in progress in the Southeast invite further inquiry about southeastern ozone control effectiveness. During the most recent time period for which data were available (1994–2003), the southeast states' AQI across MSAs rose by an average of 7.2, almost twice that of the rest of the

Table 3.1.   Modeling of Expenditure Effects on Change in State Ozone Air Quality Index 1994–2003

Effect	Model 1	Model 2	Model 3
Location	−3.4 (−1.99)**	−2.9 (−1.87)*	−3.1 (−1.96)*
Air Quality Expenditures Per Capita	−0.17 (−0.42)		0.36 (0.07)
Environmental Expenditures Per Capita		−0.05 (−3.08)**	−0.06 (−3.13)**
N =	47	46	46
R2 =	0.09	0.25	0.27

* = Significant at 0.10 level
** = Significant at 0.0 5 level

[NOTE: Model of combined total and per capita expenditures had only location and environmental expenditures per capita as significant]

United States. Further, the non-southeastern states displayed considerably greater range in increases, emphasizing the similarity of southeastern ozone circumstances.

To initiate examination of the efficiency of ozone control across regions, both aggregate and per capita expenditures for air quality among the states in 2000 were compared. The southeastern states averaged spending over 40 percent more on average ($23 million vs. $16 million) on air quality than the rest of the states, and the range of expenditures was much smaller among the southeastern states, indicating substantial consistency among southeastern states. When expenditures are adjusted for population, the southeastern states on a per capita basis spent approximately 1/3 less than other states ($2.60 vs. $3.80 per capita), with the range for the rest of the United States being over five times greater than the Southeast's. When the consideration turns to efficiency, while both groups saw the MSA AQI worsen during the period the Southeast states spent over 50 percent more per unit of change as did other states ($3.50 vs. $2.00 per person per AQI unit change). In short, it appears that the Southeast spent less per person and achieved less per person than the rest of the United States. This condition was confirmed nationally when location and results were determined to have a correlation of 0.09 at a significance level of 0.05.

State budget allocations frequently are viewed as influential upon state environmental success (Ringquist 1993). Table 3.1 shows that a southeastern location matters when overall environmental expenditures are considered alone or in combination with air quality expenditures. It is noteworthy that air quality expenditures do not appear as significant but overall environmental expenditures do. This suggests a general spillover effect of state environmental capacity but not a close tie to individual categories of pollution control program expenditures.

The importance of a southeastern location is reinforced when the states' economic conditions are employed as predictor variables along with a southeastern location (table 3.2). Whether per capita or aggregate gross state product is examined, southeastern location still matters. When state policy preferences were included in modeling, importance of location continued to be significant. However, when politics or interest group strength were employed as part of a model, neither locational importance nor the importance of interest groups or political alignment presented significance.

When economic categories of influences were combined, aggregate environmental spending per capita and southeastern location continued to matter (table 3.3). This presentation of significance of general environmental expenditures suggests that general environmental capacity may be affecting locational outcomes. When a combined economics and policy model was examined, environmental spending continued to be significant (table 3.3). A combined politics and policy model did not yield significance

Table 3.2. Modeling of Economic and Policy Alignment Effects on Change in State Ozone Air Quality Index 1994–2003

Effect	Model 4	Model 5	Model 6
Location	−3.4 (−2.01)**	−4.4 (−2.46)**	−3.8 (−1.74)*
Gross State Product	0.00 (1.26)		
Gross State Product Per Capita		196.9 (1.48)	
State Policy Liberalism per Erickson, Wright & McIver			0.31 (0.33)
N =	46	46	45
R2 =	0.12	0.14	0.09

* = Significant at 0.10 level
** = Significant at 0.0 5 level

[NOTE: Models of interest groups and political alignment presented no significance.]

Table 3.3. Modeling of Combined Categories of Effects on Change in State Ozone Air Quality Index 1994–2003

Effect	Model 7	Model 8
Location	−3.2 (−1.92)*	−1.9 (−0.82)
Environmental Expenditures Per Capita	−0.05 (−2.64)**	−0.05 (−2.46)**
Gross State Product Per Capita	71.0 (0.53)	84.0 (0.47)
State Party Alignment		0.10 (1.250
State Ideology Identification		−0.04 (−0.24)
State Policy Liberalism per Erickson, Wright & McIver		0.32 (0.24)
N =	46	45
R2 =	0.26	0.29

* = Significant at 0.10 level
** = Significant at 0.0 5 level

in any variables. The persistence across models of general environmental per capita expenditures along with a southeastern location suggests that the comparative lack of ozone progress observed in the Southeast may be related to the differences in state funding for environmental activities in general in the Southeast.

IMPLICATIONS AND INTERPRETATIONS

Ozone air pollution is an important problem for the United States and the southeastern states share that problem with the rest of the country. With a large number of growing urban areas, the Southeast presents a number of areas that do not attain the ozone national ambient air quality standard. Progress in these southeastern nonattainment areas lags that of elsewhere, in part because the seriousness of ozone pollution in the Southeast was less to begin with and the region's states had less far to go to achieve ozone attainment.

Is southeastern state ozone air quality management different from the rest of the United States? It appears to be the case. And it is different due to an overall underfunding of southeastern environmental programs rather than air quality programs alone. When ozone problems are examined across metropolitan statistical areas, the Southeast continues to be different from the rest of the nation. In the period 1994–2003 the ozone air quality index across southeastern urban areas worsened almost twice as much as other areas. In addition, southeastern states spent less per person on air quality than other states. And when the decline in air quality and the differences in state air quality expenditures are combined, southeastern states appear more inefficient in ozone management than other states. Southeastern states' ozone control is more expensive in terms of unit efficiency and achieves less than the rest of the country. The results of this analysis suggest that ozone air quality in southeastern areas is growing worse, and doing so at higher costs than in the rest of the United States.

What is the source of such differences? Put simply, it appears that institutional capacity to protect the environment at the state level matters considerably and southeastern states are falling short in that regard. This is a general capacity, not specific simply to air quality. Expenditures per capita on state environmental programs matter in ozone air quality outcomes. The southeastern states need to attend to overall environmental program capability, and if they do so are likely to harvest ozone air quality improvements. Such attempts to pursue policy-tuning appears to contribute to southeastern states lagging in ozone program control effectiveness.

This finding appears at first to contradict Ringquist's (1993) findings. He found that expenditures for water quality and air quality could not be related to outcome improvements. The findings from this ozone study suggest that it may not be appropriate to focus closely upon media-specific state environmental programs, but to attend to general institutional environmental program capacity at the state level if we wish to improve outcomes. It does not appear to be worthwhile to try to fine-tune policies, but rather rely upon broad, strong institutional capabilities to yield results. Such a situation has particular importance to southeastern states. With a bias to restrict government regulations and to closely relate expenditures to outcomes, these states are seeking to manage ozone air quality in a manner that does not appear to advance successful regulatory action.

These findings suggest that in order to make progress in ozone air quality, the southeastern states need to reevaluate their environmental policies, especially as they concern financial support. What they have been doing is not working well and is more expensive than it needs to be when compared to the rest of the United States. Not only are southeastern states alike in their resources and results, these similarities point to a shared deficiency in environmental management that, if left unaltered, will likely result in air quality degradation across the Southeast. In order to improve ozone air quality this research suggests that southeastern states must attend to building stronger overall environmental institutions and reduce attempts to fine-tune the expenditures-outcomes relationships. This will be a tall order in a region in which tight control on administrative government is an article of faith. The Southeast appears to face a choice of holding onto uniquely southern values of limited government enterprise and paying the price in declining air quality, or altering those values and behaviors to comport more closely with those of the rest of the United States, or changing such behaviors and moving toward more environmentally effective governmental practices.

REFERENCES

Bureau of Economic Analysis. 2003. Regional Accounts Data. www.bea.gov/bea/regional/gsp/action (accessed on August 7, 2005).

Environmental Commissioners of the States (ECOS). 2004. *State Environmental Expenditures*. Washington, DC: Environmental Commissioners of the States.

Erickson, R. S., G. C. Wright, and J. P. McIver. 1993. *Statehouse Democracy*. New York: Cambridge Press.

Lowery, D. 1998. *Data Table of Interest Groups*. Chapel Hill, NC: Department of Political Science.

Lowry, W. R. 1992. *The Dimensions of Federalism: State Governments and Pollution Control Policies*. Durhan, NC: Duke University Press.

Natural Resources Defense Council. 1995. Membership Profile. Washington, DC: Author.

Revez, R. L. 1997. *Foundations of Environmental Law and Policy*. New York: Foundation Press.

Ringquist, E. J. 1993. *Environmental Protection at the State Level*. Sage: London.

Rosenbaum, W. A. 2008. *Environmental Politics and Policy*. 7th ed. Washington, DC: CQ Press.

U.S. Census Bureau. 2003. *Population Estimates: ST99-3*. Washington, DC: U.S. Census Bureau.

U.S. Environmental Protection Agency. 2004. *Health and Environmental Impacts of Ground-level Ozone*. Washington, DC: U.S. EPA.

U.S. Environmental Protection Agency Office of Air Quality Planning and Standards. 1996. *Criteria Document for Revision of the Ozone National Ambient Air Quality Standard*. Research Triangle Park, NC.

———. 1997. *Regional Approaches to Improving Air Quality*. Research Triangle Park, NC.

———. 2004a. *National Air Quality and Emissions Trends Report*. Research Triangle Park, NC.

———. 2004b. *National Air Quality Index Trends*. U.S. EPA 2004. www.epa.gov/airqualitytrends (accessed on July 15, 2005).

4

Dirty Water, Clean Water

Infrastructure Funding and State Discretion in Southern States

John C. Morris

INTRODUCTION

One of the lasting legacies of the Reagan era is a movement to decentralize federal programs implemented by states. Driven by a desire to shrink the size and cost of national government, the Reagan administration pushed for legislation that would significantly strengthen state control over the design, implementation, and administration of federal programs (see Conlan 1988). States were understood to be ready and willing to assume control of these programs, and would implement the programs to best fit their particular circumstances.

Two such programs are the Clean Water State Revolving Fund program (CWSRF), created by the Water Quality Act of 1987 (P.L. 100-4), and the Drinking Water State Revolving Fund (DWSRF) program, enacted by Congress in the Safe Drinking Water Act of 1996 (P.L. 104-82). The DWSRF also contains a number of provisions that allow states to commingle both administrative structures and fund accounts for the two loan fund programs. Taken together, these two programs represent a significant shift in the roles of both states and the national government in environmental infrastructure funding. States clearly have greater administrative freedom to manage their funding programs (Morris 1997, 1994; Heilman and Johnson 1991) and have just as clearly taken different paths in program design and implementation.

The purpose of this chapter is to examine three southern states—Alabama, Mississippi, and Georgia, each of whom have fully implemented both the CWSRF and the DWSRF—to discern the different approaches each has taken in the design, implementation, and administration of these two environmental programs. Although the three states are in many ways alike, they are in many ways different; this paper seeks to explain the differences that led to the establishment of programs with very different designs, structures, authority relationships, and operations.

THE STATE REVOLVING LOAN FUND MODEL

Both federal programs follow a similar policy design scheme to achieve their respective goals. For the sake of brevity, we will address the development of the CWSRF; the DWSRF follows a very similar pattern.

One of the main desires of the Reagan administration was to reduce federal spending. Faced with significant budget deficits, a working group convened at U.S. Environmental Protection Agency (USEPA) in 1984 to discuss alternative means to fund wastewater treatment projects in the state. The Clean Water Act of 1972 had made the federal government the major source of funds for wastewater infrastructure needs through the use of programs for construction grants. The USEPA group was charged with finding alternative mechanisms for funding; after much discussion they recommended a revolving loan model (EPA 1984). This proposal was accepted and written into the 1987 act that created the CWSRF.

The program originally called for federal grants to states over a period of seven years (through FY 1994),[1] to be allocated through a grant formula process (see Dilger 1986). In order to receive an annual grant (called a capitalization grant) states were required to provide at least 20 percent in the form of matching funds for each grant. The funds would then be placed in an account from which the state would make loans to eligible communities for wastewater needs. In turn, the communities would pay back the loan principle with interest for the life of the loan (typically twenty years, with state-to-state variation). The returned principle, with its accompanying interest, would then be re-loaned (revolved) in the form of new loans to other eligible communities. As long as the interest rate charged by states remained above the prevailing rate of inflation, the funds would continue to grow in perpetuity; states would ultimately achieve self-sufficiency to be able to meet all their future wastewater needs. Federal funding could then cease, thus reducing federal budget requirements.

Because federal funds were tight, congressional policy makers knew the authorized amounts available for the CWSRF program would be insufficient to meet the current wastewater needs in the states, much less future needs. States were thus encouraged to leverage their available funds to create additional monies to capitalize their funds and to contribute more than the minimum 20 percent match. While leveraging mechanisms are many and complex, they all have the effect of increasing the cost of a loan (through higher interest rates) for recipient communities (see Holcombe 1992). As a result, few states chose to leverage (see Travis, Morris, and Morris 2004); fewer still have contributed more than the minimum 20 percent. Although the initial authorization expired in 1994, Congress has consistently authorized (and appropriated) additional funds to further capitalize state loan pools. A final point to note is that the 1987 legislation required states to use the funds to serve three specific classes of communities: those with significant compliance needs (a specific environmental need), small communities (with populations less than 5,000), and financially at-risk (poor) communities. For a variety of reasons detailed elsewhere (see Morris 1999; 1997; 1994; see also GAO 1992), the CWSRF has been less than successful at meeting these specific needs.

The DWSRF program is similar in its structure and came about as a result of the popularity of the CWSRF program with states. Congress provided additional state discretion, however, by allowing states to either use their existing CWSRF structures or create separate structures for DWSRF implementation, and to combine the financial structures of the two programs. Fund accounting and auditing processes are conducted on the two as separate programs, but monies can be freely commingled and transferred between one program and the other to meet specific state needs. In a survey of state DWSRF coordinators conducted in 1997, all but four percent of respondents report that they made use of at least part of their CWSRF structure for the DWSRF program; 47 percent used the same structure.

A final point to note about the revolving loan fund model is that it represents a significant change in the "who pays" part of the policy equation. Under the old construction grants program, communities could receive a grant of money for up to 80 percent of the construction costs of the new facility. Often the state would make available other sums of federal program money (such as Community Development Block Grant funds) or state funds to supplement the base grant, meaning that the costs borne by the individual community were relatively small and could be covered through user fees, bond sales, or some other mechanism. With the advent of a loan program, however, communities are now facing responsibility for an appreciably higher portion of the costs of construction, often reaching 80 percent or more. This higher burden means that communities that are able to issue bonds must commit a significantly higher portion of their bond capacity to meet wastewater needs or pass on the additional costs in the form of substantially higher user fees. While the increase in user fees in a larger community might not impose significant hardship on citizens, smaller communities have fewer users to share in the costs. Anecdotal evidence from around the nation suggests that some small communities have seen their monthly sewer bill quadruple in response to the increased costs of a CWSRF loan. These reports have given many communities (and states) pause as they consider ways to lessen the financial burden of clean water, particularly in small and financially disadvantaged communities.

THE CASE STUDIES

The three states chosen for this analysis are contiguous, Deep South states. Table 4.1 presents a series of dimensions across which to compare the three cases. The first eight items in the table provide a look at a series of state political and environmental indicators. The first item, state population, is drawn from U.S. Census data. To describe political culture we employ the measure developed by Sharkansky (1969). Although Sharkansky's index is fraught with problems when used to compare states across a range of political cultures (see Clynch 1972; Schlitz and Rainey 1978; Savage 1981), it is useful to distinguish differences between states with very similar political cultures. While all three states are classified as predominately "traditional" by Elazar (1972), the Sharkansky index suggests that Mississippi is a "pure" traditionalistic state, while Alabama and Georgia show evidence of a small "individualistic" component. This is further supported by Elazar's substate analysis, in which he concludes that Mississippi is a "traditionalistic dominant" state, while both Alabama and Georgia are "TI" states ("traditionalistic dominant; strong individualistic strain;" see Elazar 1984, 136). The government ideology indicator is an index drawn from Berry et al. (1998); lower values suggest an ideology more on the conservative end of the (0–100) scale. All three states have part-time legislatures. Likewise, the measure of interparty competition (drawn from 1998 U.S. Census data; Soss et al. 2001) is a measure of the vitality of political competition within a state legislature; higher values suggest greater competition. Finally, drawn from Bowman and Kearney (1988) is a summary of a state's relative administrative capacity across four areas: staffing and spending, accountability and information management, representation, and executive centralization. Larger (positive) numbers indicate relatively higher levels of administrative capacity. The next two variables are measures of the vitality of environmental policy in the state. The first measure, drawn from Davis and Lester (1989), is a general measure of state commitment to environmental quality, while the final measure (Lester and Lombard 1990) is a more targeted measure of state commitment to water quality. Where possible, all data are drawn from the period 1985–1990 to best match the implementation period of the CWSRF program.

The next set of indicators compares dimensions of the CWSRF program. The first figure indicates the year in which the state received its first CWSRF capitalization grant, followed by the year in which the state made its first CWSRF loan. The next dimension is the availability of nonfederal

Table 4.1. Pertinent State Information

Indicator	Alabama	Georgia	Mississippi	National Mean (Range)
General State Indicators				
Population (1988, in millions)	4.127	6.401	2.627	4.67 (.47–28.16)
Political Culture (from Sharkansky)	8.57	8.80	9.0	5.28 (1.0–9.0)
Government Ideology	30.86	77.0	25.88	39.77 (1.25–93.88)
Legislature (part-time or full-time)	Part-time	Part-time	Part-time	— (—)
Interparty Competition	.67	.80	.63	.741 (.30–.97)
State Administrative Capacity	–3.33	–1.66	–3.17	–.397 (–3.33–4.28)
Commitment to Environmental Protection	10	25	15	28.84 (10–47)
Commitment to Clean Water	16	26	14	29.78 (1–50)
Wastewater Indicators				
Date of 1st CWSRF Cap Grant (date of 1st loan)	1989 (1990)	1988 (1988)	1989 (1990)	1989 (1988-1990)
Nonfederal CWSRF Alternatives (1991, in millions)	0	1.46	0	.5 (0–500)
CWSRF Needs (1988. in millions)	547	615	410	886 (.202–5,257.0)
Drinking Water Indicators				
Date of 1st DWSRF Cap Grant (date of 1st loan)	1998 (1998)	1997 (1997)	1997 (1997)	1997 (1997–1999)
DW Needs (1999, in millions)	677.2	1,583.1	981.8	1,132.5 (115.8–1,956.6)
Same/Different Structure for CWSRF/DWSRF	nearly identical	nearly identical	nearly identical	—

wastewater funding program funds, and the third figure indicates each state's 1988 wastewater needs as reported by EPA (1989). The final set of figures represent pertinent data on each state's DWSRF program, including the dates of the first capitalization grant and first DWSRF loan, the state's drinking water infrastructure needs (EPA 2001), and whether the state used the same or a different structure for the DWSRF program. Note that we do not include nonfederal alternatives for drinking water needs; many states use portions of their Community Development Block Grant (CDBG) funds or USDA Rural Development funds (or other similar programs) to fund drinking water infrastructure, and no reliable data are available to compare relative state effort.

In short, the dimensions in the table suggest that the three states are more alike than they are different, particularly when compared to the means and ranges for all fifty states. While populations differ, they all fall into the same quintile when compared to all states. All have very similar political cultures, and all three fall below the mean on the state administrative capacity scale. All three states fall below the national means on both environmental scales; two fall significantly below the mean. Two of the three states have no nonfederal wastewater funding alternative, and all have wastewater needs near the national median. Two of the three states received their first DWSRF capitalization grant the first year they were available, and all three employed the existing CWSRF structure for their DWSRF programs. In spite of these similarities, all three states took very different paths in the development of their CWSRF programs, and thus their DWSRF programs as well. The following sections present a more in-depth look at each state and the implementation and design choices each made in both the CWSRF and DWSRF programs, and how those choices translate into program operation.

Mississippi

Mississippi is the smallest of the three states, and the state with the smallest wastewater needs. It is also the state that is the most traditionalistic, the most conservative, the least politically competitive, and the least committed to water policy. The CWSRF program is administratively located in the state's Department of Environmental Quality (DEQ) and is identical to the structure used under the Construction Grants program. The Wastewater Branch is dominated by civil engineers and has traditionally placed engineering concerns ahead of financial, political, or environmental concerns.

At the time CWSRF implementation began, Mississippi was experiencing an economic downturn, and state budgets were tight. Indeed, the state legislature was unable to budget for the required matching funds, and DEQ officials seemed unwilling to expend political capital with the legislature to force the issue. The following year the legislature did appropriate matching funds, spurred in part by representatives and senators from several districts containing communities with desperate infrastructure needs (located in growth areas in the northwest and coastal regions of the state). One DEQ official also noted that there was a great deal of internal discussion as to whether the demise of the old construction grants program was really imminent; Mississippi had been a significant beneficiary of the program and, DEQ officials thought, the new CWSRF program would clearly place costs on communities they would be unable to bear. When new construction grants were not made available, DEQ began to join the voices requesting state funding for the required match.

Both the legislature and DEQ also had real questions as to whether demand for loans would be high enough to warrant leveraging, and thus this issue was never seriously considered. It is also likely that the state would have been ill-equipped to address the significant financial expertise required by leveraging, as there is a significant institutional protectiveness asserted by DEQ. Lacking the appropriate expertise in-house, DEQ would be forced to seek such assistance outside the agency and thus surrender a considerable amount of program control to another entity. The decision was therefore made to forego leveraging and to run the program within the Wastewater Branch as a straight loan program.

The implementation of the DWSRF raised the control issue again in 1997. At this point the state was in slightly better financial shape and was able to secure its first DWSRF capitalization grant in the first year they were available. However, administrative responsibility for drinking water in Mississippi is housed in the Department of Health (DoH), and something of an interagency squabble developed between DoH and DEQ for control over the DWSRF. The DEQ was concerned that to cede control over the DWSRF would mean that some amount of control over the CWSRF would also be lost, since the new program allowed for the commingling of program funds. They also asserted that auditing and accountability reviews would be made more difficult because of the split responsibilities between agencies. On the other hand, the DoH had no interest in giving up its historical control over drinking water. They argued that they already had the testing, engineering, and scientific infrastructure in place to best administer the program, and that they also had developed very close relationships with the drinking water policy community in the state, including the politically powerful Mississippi Rural Water Association (MRWA). Mississippi has over 1,400 drinking water systems, many of them small, private water associations that often see government action of any kind as anathema to individual rights. The MRWA had been a very effective interface between the DoH and the water systems, often facilitating communication in both directions. One DoH engineer tells an entertaining story of being forced to leave a meeting of angry water association customers by jumping through a church window. The MRWA representative also in attendance was able to calm the crowd enough to allow the engineer to return (through the door) to continue the meeting.

The compromise finally reached was to split responsibility for the DWSRF between the two. The DoH gained programmatic control of the program, which allowed the agency to use its existing connections to develop loan requests, engineering and health needs, and to monitor construction and upgrades. The DEQ received budgetary control, which not only gave it additional funding responsibility, but also allowed it to safely commingle funds as necessary to achieve financial goals. The heads of relevant offices in each agency report a solid and productive working relationship, and neither side to date has seen fit to seek legislative or judicial intervention to solve disputes.

In sum, the Mississippi story is of a state that faced both political and financial hurdles to implementation and was significantly limited in its choices as a result. Both programs are run as traditional state funding programs, with only minimal sharing of authority between agencies. State budget woes have delayed or prevented the state receiving its capitalization grants on occasion, and there are significant needs, particularly in the wastewater area, that have gone unmet because of the higher cost of CWSRF assistance (as compared to the construction grants program). The fund has remained viable, but barely; as a way to subsidize the costs of a loan, the state has kept loan interest rates very low. This has meant that loan rates have barely kept ahead of the rate of inflation. Mississippi officials report that they have no future plans to engage in leveraging, and to date they have yet to make any transfers between their CWSRF and DWSRF accounts.

Alabama

Alabama is the middle state of the three, in both location and population. The state scores the lowest of the three on the state administrative capacity index (indeed, it scores the lowest of any state in the nation) and also has the lowest combined environmental scores. Both the CWSRF and the DWSRF programs are administratively located in the Alabama Department of Environmental Management (ADEM), in the same agency that had responsibility for the old construction grants program.

Both Elazar (1972) and Sharkansky (1969) describe Alabama's political culture as traditionalistic/individualistic. While the patterns of interpersonal relationships tend to favor the traditionalistic conception, Alabamians often extol the virtues of a *laissez faire* market. The "engine" of politics in Alabama is interpersonal relationships. Decisions are made as agreements between gentlemen and often outside the patterns of formal authority and communication found in bureaucracies. The political culture in Alabama thus places a premium on the sovereignty of the market and decision making through personal and political relationships and contacts.

Like Mississippi, Alabama was suffering from a weak economy in the late 1980s, and the state legislature balked at the request to appropriate funds for the required match. When the U.S. Congress passed the Water Quality Act in 1987, Alabama was caught in a period of uncertainty, since nearly all of Alabama's water quality infrastructure resources came from the federal government via the construction grants program. A private sector investment banker, sensing an opportunity to develop a business opportunity, used his contacts in the state legislature to shape and enact the enabling legislation for the Alabama CWSRF program. Working with select members of both houses of the legislature, the investment banker drafted the legislation, lobbied members of the legislature, and became the force behind the bill. As one state official said, "[he] selected Alabama; Alabama didn't select [him]."

The legislation was seemingly simple and innocuous. It created a "shell" organization, the Alabama Water Pollution Control Authority (AWPCA), but did not authorize any funds for the new authority. However, the legislation did give the AWPCA the authority to issue state general revenue bonds, make loans to communities, and assume administrative control of the CWSRF program. The AWPCA consists of five formal members, all of whom are public officials—the governor, the lieutenant governor, the Speaker of the House, the state finance director, and the director of the Alabama Department of Environmental Management (ADEM). The Authority is constructed in such a way that the governor always controls a majority of members, since the finance director and the director of ADEM are gubernatorial appointees. The enabling legislation gave AWPCA all formal program authority, including the authority to make decisions about the allocation of loans. The Facilities Construction Section of ADEM, formerly the Wastewater Treatment Works (WTW) construction grants program office, provides administrative support in terms of maintaining a priority list of applicants, reviewing applications and plans, and overseeing general program administration.

The major formal actor in the Alabama CWSRF is the Alabama Water Pollution Control Authority (AWPCA). As discussed previously, nearly all formal programmatic authority in the CWSRF program rests with this body. The Authority has no funding or staff and thus relies on formal contracts with four different private sector actors—a trustee bank, a financial advisor, a bond attorney, and a bond underwriter—to operate the program. However, there is a significant difference between the formal authority specified in the enabling legislation and the actual exercise of authority in the program. Although the AWPCA is given nearly all formal program authority under the legislation, two actors—the financial advisor and a state official in ADEM—make nearly all program decisions. Of these two, the most important actor is the private sector financial advisor. In many respects, the financial advisor *is* the CWSRF program in Alabama. Although given nearly all formal program authority, the real role of the AWPCA is reduced to what one state official described as "benign neglect." The financial advisor is, without a doubt, the preeminent actor in the Alabama CWSRF.

Alabama engages in an aggressively leveraged financial structure for the CWSRF program. In fact, Alabama floats bond issues every year to meet the required 20 percent state match for the federal capitalization grant, since the state legislature has never appropriated funds for the state match. The financial structure used by Alabama is complex, and involves several revenue streams, reserve funds, and bond repayment funds. The federal capitalization grant is used to provide administrative costs and to capitalize a bond repayment account. The investment banker creates "pools" of applicants that are ready to proceed with the loan process and attempts to secure a favorable bond rating for the pool, negotiated with private bond raters such as Standard and Poor's or Moody's. If the pool contains some applicants with marginal bond ratings such that a favorable bond rating may be uncertain, the investment banker may recruit "elephants"—large or financially solvent communities with high bond ratings—to enter the bond pool. This strategy has two important effects. First, the state can receive more favorable terms for the bonds by drawing on the credit rating of the financially sound communities. By raising the bond rating of the pool, the overhead costs of the bond issue (insurance costs, etc.) are reduced. Second, the inclusion of "elephants" may compromise the priority list, since communities are offered assistance based on their financial condition rather than their relative environmental need. This is not to say that the "elephants" do not appear on the priority list or do not have legitimate water quality needs, but rather those communities with greater environmental needs (and a lesser financial status) may be denied assistance. Thus the financial status of a community may take precedence over its environmental need.

In spite of the apparent efforts to maintain the financial health of the CWSRF program, state officials expressed some doubt about the long-term viability of the fund. As one state official said, "we've sold out the future" for present needs. Although the state official thus agreed that aggressive leveraging has long-term negative consequences for the CWSRF program, he cited ambiguity in both the Water Quality Act and in USEPA documents about the meaning of "perpetuity." "Nowhere in the CWSRF program—in the EPA regulations, initial guidance, the [national] legislation, or other documents—does it give a definition of 'perpetual.' We're not sure what this means," he said.

Implementation of the DWSRF program followed a much simpler path, although the timing of the national legislation did not coincide well with the state legislature's session, requiring the delay of a year before Alabama received its first capitalization grant. ADEM also had previous responsibility for drinking water, so the only issue to be decided was how to allocate financial responsibility for the new program. Although Alabama already had a successful leveraging arrangement in place for the CWSRF, the legislature opted to create a separate board, the Alabama Drinking Water Finance Authority (ADWFA). The structure and membership of the authority is identical to the AWPCA, and in fact the two are separate in name only. The only material difference is that a different branch in ADEM is responsible for working with DWSRF applicant communities, developing priority lists, and so on. Indeed, the same private sector financial advisor runs both programs, although, as of this writing, Alabama has not commingled funds in any year since the inception of the DWSRF.

In sum, Alabama took a very different path than did Mississippi. The political culture of Alabama created an opportunity for a private investment banker to seize an opportunity made available by the inability of the state to act, and the result is a program that exhibits very little public control over these environmental resources. The Alabama SRF programs are the most heavily privatized of any in the country (Morris 1994, 1997), although it is also useful to understand the difference between "formal" and "informal" program authority. While the enabling legislation for both programs created public authorities charged with program responsibility and provided administrative support roles to ADEM, nearly all program decision activity takes place outside this formal structure.

Georgia

Georgia is the largest of the three states discussed here, and the most different across most of the indicators presented in table 4.1. Georgia has the most liberal government of the three cases, the highest level of interparty competition, the highest level of administrative capacity (although still under the national mean), the highest commitment to environmental and water protection, and the largest needs for both wastewater and drinking water. of the three states. Georgia was also an early implementer of both the CWSRF and the DWSRF programs, receiving capitalization grants in their first years of availability. Finally, Georgia is the only one of the three that operates a nonfederal funding source for wastewater needs. Program responsibility for both the CWSRF and DWSRF programs rest with the Department of Natural Resources (DNR) and the Georgia Environmental Facilities Authority (GEFA).

Georgia was among the first states to adopt the CWSRF program. Indeed, the enabling legislation to create the CWSRF program in Georgia was passed by the state legislature several months before the passage of the Water Quality Act in 1987. State officials in Georgia cited a very good working relationship with the USEPA's Region IV office in Atlanta as a major factor in the early passage of the legislation. Region IV personnel acted as intermediaries between USEPA headquarters and state officials in Georgia. The result was the passage of a short (three paragraphs), somewhat vague bill that gave authority to state officials to create and administer a revolving loan fund program, pending the outcome and details of the national legislation.

One of the key points cited by state officials was that passage of the legislation in Georgia was made smoother by the fact that the legislature did not need to appropriate additional funds to the CWSRF program. As mentioned previously, Georgia operates a state-funded infrastructure-financing program under the auspices of GEFA. A provision of the Water Quality Act allows states to use "in-kind" matches for the federal capitalization grant, and Georgia received the approval of USEPA to apply state loans made through GEFA to the match requirement. A more practical reason was a perception on the part of state officials that if the legislature needed to appropriate money directly to the CWSRF program, their response would be to take money away from GEFA and give it to the CWSRF program. State officials in both the DNR and GEFA agreed that such a move would not be in the best interests of either the state or the environmental goals of the state. State officials, with the assistance of the Region IV office, negotiated the use of an in-kind match that allowed Georgia to receive the federal capitalization grant without the expenditure of additional state funds. GEFA does have authority to issue general obligation bonds.

Although early demand for CWSRF assistance was substantial, Georgia has never operated a leveraged program. In the early years of the program, the investment banking industry, both local and national, tried very hard to convince state CWSRF officials to leverage the program. State officials resisted the pressure for two reasons. First, state officials felt that while early demand was high (three or four times as much as the available program resources), the combination of CWSRF loans and GEFA loans would cover the demand. Second, state officials were very concerned about the effects of leveraging on the long-term solvency and perpetuity of the fund. State officials were convinced that while leveraging would provide a greater pool of money in the near-term, the long-term effects of leveraging would significantly degrade the value of the fund over time. As one state official said, "we felt that aggressive leveraging is a sure way to *not* have a fund in perpetuity. We came to the conclusion that it was not in the best interests of the state to leverage."

Until July 1994, the CWSRF program in Georgia was administered jointly by the Environmental Protection Division (EPD) of the Georgia

Department of Natural Resources and GEFA. Under the old system, EPD assumed responsibility for many of the programmatic decisions made in the CWSRF program. EPD processed applications for assistance, reviewed design plans and specifications, maintained the priority list, performed environmental assessments, and tracked loan repayments. GEFA officials assumed mostly coordinative activities, including a financial review of applicants, monitoring loan accounts, and preparing loan materials for execution. The director of the EPD and the commissioner of GEFA signed loan agreements for the state.

A new administrative mechanism was implemented in July 1994. The most obvious change in the CWSRF structure is that programmatic authority for the CWSRF program shifted from EPD to GEFA. Accompanying the shift in authority was a change in the duties of both EPD and GEFA. EPD retained responsibility for the environmental/technical aspects of the program (design review, environmental impact assessments, etc.) and tracked loan repayments for the program. The changes for GEFA were more substantial. As a means of consolidating program authority, GEFA assumed program authority for both the state-funded loan program and the CWSRF program. According to a state official, the primary force behind the reorganization was a desire to centralize funding authority. To better coordinate the funding decisions of both GEFA and the CWSRF program, a decision was made to transfer the program authority to GEFA. In many ways, this shift is representative of the inherent tensions between the environmental goals and the financial goals within the CWSRF program. As one EPD official indicated, the financial requirements of the CWSRF program fundamentally alter the administrative approach required to administer a successful program. "We're not a bank," the official said, "but we often have to act like one." By transferring program authority to GEFA, the CWSRF program gains an experienced and successful state agency to administer the financial portions of the program and to allow in-house coordination of the CWSRF program with the state-funded program administered by GEFA.

An additional feature of the interagency agreement between GEFA and EPD is the availability of EPD's technical and administrative law to GEFA in its CWSRF activities. Rather than duplicate the administrative and technical authority of EPD, GEFA shares the necessary authority with EPD. EPD still maintains much of the design and technical authority in the program, but GEFA, through its assumption of program authority, can draw on EPD's technical and administrative law to manage the CWSRF program more efficiently.

The structure of Georgia's DWSRF is nearly identical to that of its CWSRF. As with the CWSRF, Georgia was an early implementer of the DWSRF, and employed the identical structure used in the older program. As before,

GEFA assumes primary responsibility for the financial elements of the program, while EPD assumes responsibility for the permitting and engineering processes. Like the other two states, Georgia has chosen not to commingle their CWSRF and DWSRF funds, maintaining separate accounts and funding sources for each program. The viability and experience of GEFA's infrastructure financing experience, coupled with administrative and engineering expertise colocated in the same agency, meant that the implementation of Georgia's DWSRF program was accomplished with little fanfare.

In sum, Georgia's experience is very different than our two previous cases. Not only was Georgia much more proactive in the creation of the programs, they were the only one of the three with an existing state program prior to CWSRF implementation (consistent with their higher scores on the environmental indicators presented in table 4.1). Program design and implementation proceeded without internal administrative squabbling, and implementation was achieved through the exclusive use of state government officials. Georgia's more conservative fiscal approach mirrors that of Mississippi, but unlike Mississippi, GEFA has the in-house expertise to accomplish leveraging if they were to so choose.

Comparison of the Cases

Mississippi, Alabama, and Georgia represent three very different responses to the implementation and operation of the two federal revolving loan programs. We now turn to a comparison of the three cases across three different dimensions: program design and development, program structure and authority, and program operation.

Program Design and Development

Georgia was not only an early implementer of both programs, they had state legislation in place to create their CWSRF before the federal legislation creating the CWSRF had been passed and signed into law. Georgia's interest in water issues is also evidenced by the previous existence of a state-funded program to fund wastewater needs in the state. Georgia credits a close working relationship with Region IV USEPA officials for their ability to implement early; this is likely helped by the fact that the USEPA's Region IV offices are also located in Atlanta. On the other hand, both Alabama and Mississippi were initially constrained by a lack of state resources to meet the program match requirements. Two different forces came into play: in Mississippi, the pressures of growth and subsequent pressure by local officials finally forced the legislature to act, while in Alabama a private investment banker stepped into the vacuum of inaction and used his personal influence to create a market opportunity.

Program Structure and Authority

Mississippi and Georgia both locate their programs wholly in the public sector. In each case, public agencies make all authoritative decisions regarding program design and the distribution of program resources. Georgia has split the environmental components of the program into different agencies, while Mississippi, with the exception of a role for the Department of Health in the DWSRF, has retained the programs in a single public agency. Alabama, on the other hand, has effectively privatized the decision-making function of their programs by turning over a substantial portion of program authority to a private sector consultant. While many states make use of program consultants (see Morris 1994), an important difference is that Alabama's consultant plays a significant role in decisions regarding the distribution of program resources. More importantly, this program authority is highly personalized and informal; there are no laws, rules, or contracts specifying the role of the consultant. In all three states, a traditional state agency (environmental or health) maintains contacts with local officials, handles applications for assistance, sets the priority lists, and approves the engineering associated with each project.

Program Operation

All three states operate their programs as direct loan programs, meaning they make new loans directly to applicant communities. Two of the three states provide interest rate reductions for small or financially at-risk communities; Alabama influences its interest rates by structuring each bond pool to achieve the highest possible bond rating and thus the lowest interest rate. In all three states, applications for loans are approved through a formal process; in Mississippi this involves a process internal to DEQ (and with DoH, in the case of DWSRF loans). Alabama and Georgia both require the approval of both an environmental agency and a separate financing authority, although in Alabama such authority is largely a "rubber stamp"— actual funding decisions are made by the private consultant through the creation of each bond pool.

Finally, while state coordinators all report that demand for assistance is steady and water needs are unmet, none of the three states have commingled program funds, and only Alabama has engaged in leveraging. The reasons for the latter are several. First, Mississippi lacks the in-house expertise to engage in leveraging, and state officials are concerned about the additional costs of leveraging that would be passed on to applicant communities in the form of higher interest rates. Georgia has the same concern about interest rates, but clearly has the required expertise in GEFA. Alabama, on the other hand, has leveraged every year since the inception of the CWSRF. The explanation lies in the relationship between the state entities and the financial consultant. Since the financial consultant works without a contract, he is not paid a flat fee for his services. Rather, his income in this program is derived from his role as a financial actor—he receives compensation as a bond broker and receives a fee from the formation of a bond pool and the sale of bonds. In addition, the financial consultant gains valuable access to local government officials to market additional financial services for other municipal needs.

Political Culture, Commitment to Clean Water, and Administrative Capacity

The case studies illustrate the combined influences of these three factors. In Georgia, the state with the highest administrative capacity of the three, administrators of the program were able to draw on the expertise already present in the state-funded loan program to design and operate both the CWSRF and DWSRF programs. While highly traditionalistic, the administrative expertise of the program administrators meant that integration of the two federal revolving fund programs was accomplished without drama. Coupled with the highest scores of the three states in terms of their commitment to water quality, it is not surprising to find that Georgia not only had a state-funded program that preceded the federal programs, but employed that expertise and experience to design and implement the two federal programs.

Alabama, on the other hand, is a state with a very different circumstance. The political culture score of the state leans the most toward the "individualistic" side of the three, and it suffers with the lowest administrative capacity of the three. Lacking the expertise to design and administer the federal revolving fund programs, Alabama ultimately turned to the private sector for assistance. Because individualistic cultures tend to favor private interests, it is no surprise that the state seized the offer of help from a private sector actor. Moreover, while there is public-sector control over the program at least in theory, the tremendous amount of discretion and control over public resources given to the private-sector actor (see Morris 1997) is a direct result of the combination of a political culture that encourages individual benefit and a lack of expertise and capability in the relevant state agencies. Alabama's lack of commitment to the environment in general and to water quality in particular helps explain their willingness to cede control of the program to a non-state actor.

Finally, Mississippi's circumstances are also a product of these same factors. As a highly traditionalistic state (indeed, the most traditionalistic state in the nation), government control over public resources is absolute. Although the state also scores very poorly on the administrative capacity index, the dominant political culture prevents the state from seeking expertise outside of state government. Moreover, there is little incentive to do more than the minimum necessary with the program—a lack of state funds, coupled with a lack of interest in water quality, means that the state will do the minimum necessary to meet the national policy imperatives.

CONCLUSION

The case studies presented here attest to the significant differences in the design, implementation, and operation of the two revolving loan fund programs designed to meet state water infrastructure needs. In one sense, the programs have fulfilled their promise of greater state discretion, as illustrated by the wide variation in which these three states have approached their respective choices. The development of each state's program is clearly a product of the specific circumstances of the individual state, shaped by the unique forces present at that point in time. Although all three states fall below the national means on the two environmental commitment scores presented, each state is seeking to maximize its benefit from the resources available, within its contextual bounds.

These three cases also illustrate the significant variation that exists between states that appear, at first blush, to be more alike than different. The themes presented in these cases—political culture, state administrative and fiscal resources, and commitment to clean water—are not unique to these three states, but the means by which these themes manifest themselves in the policy choices of these states *are* unique. While all three states were constrained at some level by fiscal shortages, they chose very different paths to address the issue. Georgia made use of an existing state program, while Mississippi chose a minimalist approach (do as little as possible, and only then after a concerted effort to lobby the state legislature). On the other hand, Alabama did not even attempt to secure fiscal resources from the legislature; rather, they seized the offer from a private sector consultant to raise the state matching funds through an aggressive leveraging scheme. In each case, the political culture of the state also helped define the possible courses of action for each state, as did the administrative capacity of state government. Coupled with differing levels of commitment to water quality, these factors led to unique programs in each case.

Three broad conclusions can be reached from this work. First, resources matter, in that their availability determines the policy options available to states. Whether fiscal, administrative, or political, resource levels either create or limit state options in program design and implementation. It is reasonable to suggest that had Georgia not had its own state-run funding program, or Alabama had more politically adept leadership in ADEM, these three states may well have ended up with very similar programs. Instead, resource levels become an important explanator of the programmatic and administrative differences between states.

Second, it is difficult to speak in generalities about the implementation of federal environmental programs in southern states. Though many of the more standard comparative state indicators suggest these three states should be very much alike, their respective programs are more different than they are alike. In short, indicators alone are often not sensitive enough (nor comprehensive enough) to capture subtle differences in state circumstances. Each state's policy choices are the product of a unique set of conditions and factors present in the state at that time, and thus reflect those conditions and factors. Likewise, the prevailing political culture in each state helps explain the choices made in each state. While the differences in political culture between states may seem small, even small differences can have significant effects on program design and implementation (Breaux and Morris 2001).

Finally, the revolving fund programs have in many ways fulfilled the promise of increased state discretion in the implementation of national programs. As these three cases illustrate, states truly are "policy laboratories" in which states work within a national policy framework to achieve national policy goals, but are free to mold the specifics of their policy choices to their own requirements and circumstances. The story of the implementation of the revolving fund programs in these three states is one of innovation within the constraints of the unique circumstances of each state. Faced with the same policy imperative but in different environments, each state created a program that fit the political, fiscal, administrative, and social environments of each. In the end, the goal is the same; however, it manifests itself in very different ways in different states.

NOTE

1. A bill passed by Congress in 2004 further extended authorization for the two SRF programs over five years, allocating a total of $20 billion to the CWSRF and $15 billion to the DWSRF program (Ichnioski 2004).

REFERENCES

Berry, W. D., E. Ringquist, R. Fording, and R. Hanson. 1998. Measuring Citizen and Government Ideology in the American States, 1960–1993. *American Journal of Politics* 42:327–48.

Bowman, A. O'M., and R. C. Kearney. 1988. Dimensions of State Government Capability. *Western Political Quarterly* 41:341–62.

Breaux, D. A., and J. C. Morris. 2001. Assessing the Utility of Political culture in Explaining Interstate Variation in Policy Outcomes: The Case of TANF. Paper presented at the 73rd Annual Meetings of the Southern Political Science Association, Atlanta, GA, November.

Clynch, E. J. 1972. A Critique of Ira Sharkansky's "The Utility of Elazar's Political Culture." *Polity* 5:139–41.

Conlan, T. 1988. *New Federalism: Intergovernmental Reform from Nixon to Reagan.* Washington, DC: Brookings.

Davis, C. E., and J. P. Lester. 1989. Federalism and Environmental Policy. In *Environmental Politics and Policy: Theories and Evidence*, edited by J. P. Lester, 57–84. Durham, NC: Duke University Press.

Dilger, R. J. 1986. Grantsmanship, Formulamanship, and Other Allocational Principles: Wastewater Treatment Grants. In *American Intergovernmental Relations Today: Perspectives and Controversies*, edited by R. J. Dilger. Englewood Cliffs, NJ: Prentice Hall.

Elazar, D. 1972. *American Federalism: A View from the States.* 2nd ed. New York: Thomas Crowell.

Elazar, D. 1984. *American Federalism: A View from the States.* 3rd ed. New York: Harper Row.

Environmental Protection Agency (EPA). 1984. Study of the Future Federal Role in Municipal Wastewater Treatment. Washington, DC: EPA, Office of Municipal Pollution Control.

———. 1989. 1988 Needs Survey Report to Congress: Assessment of Needed Publicly Owned Wastewater Treatment Facilities in the United States. Washington, DC: EPA, Office of Municipal Pollution Control. (EPA 430/09-89-001).

———. 2001. *Drinking Water Infrastructure Needs Survey: Second Report to Congress.* Washington, DC: EPA, Office of Water. (EPA 816-R-01-004).

General Accounting Office (GAO). 1992. *Water Pollution: State Revolving Funds Insufficient to Meet Wastewater Treatment Needs.* Washington, DC: Author. GAO/RECD-92-35.

Heilman, J. G., and G. W. Johnson. 1991. *State Revolving Loan Funds: Analysis of Institutional Arrangement and Distributive Consequences.* Auburn University, AL: Final Report submitted to the U.S. Geological Survey, Department of the Interior.

Holcombe, R. G. 1992. Revolving Fund Finance: The Case of Wastewater Treatment. *Public Budgeting & Finance* 12:50–65.

Ichnioski, T. 2004. Water: Panel Authorizes Increase for Revolving Funds. *Engineering News Record* 253, 1 (July 5, 2004):11.

Lester, J. P., and E. N. Lombard. 1990. The Comparative Analysis of State Environmental Policy. *Natural Resources Journal* 30:301–19.

Morris, J. C. 1994. *Privatization and Environmental Policy: An Examination of the Distributive Consequences of Private Sector Activity in State Revolving Funds.* Unpublished doctoral dissertation, Auburn University, AL.

———. 1997. The Distributional Impacts of Privatization in National Water Quality Policy. *Journal of Politics* 59, 1:56–72.

———. 1999. State Implementation of National Water Quality Policy: Policy Theory, Policy Streams, and (Un)Intended Consequences in Water Quality. *Southeastern Political Review* 27, 3:317–30.

Savage, R. L. 1981. Looking for Political Subcultures: A Critique of the Rummage-Sale Approach. *Western Political Quarterly* 43:331–36.

Schlitz, T. D., and R. L. Rainey. 1978. The Geographical Distribution of Elazar's Political Subcultures among the Mass Population: A Research Note. *Western Political Quarterly* 31:410–15.

Sharkansky, I. 1969. The Utility of Elazar's Political Culture: A Research Note. *Polity* 2:66–83.

Soss, J., S. F. Schram, T. P. Vartanian, and Erin O'Brien. 2001. Setting the Terms of Relief: Explaining State Policy Choices in the Devolution Revolution. *American Journal of Political Science* 45:378–95.

Travis, R., J. C. Morris, and E. D. Morris. 2004. State Implementation of Federal Environmental Policy: Explaining Leveraging in the Clean Water State Revolving Fund. *Policy Studies Journal* 32:461–80.

5

Water Wars in the South

Considering the ACT and ACF Interstate Compacts

James Newman

INTRODUCTION

Since the early 1990s, the states of Alabama, Georgia, and Florida have argued over access to fresh water in the Apalachicola-Chattahoochee-Flint (ACF) River Basin and the adjacent Alabama-Coosa-Tallapoosa (ACT) River Basin (Dellapenna 2002; Hart 2003; Hyatt 2004). In hopes of developing a solution to these arguments, in 1997 the states created the ACT and ACF interstate compacts (Shelton 2003; Stephenson 2000).

Arguments over water allocation are uncommon in eastern regions of the country. Residents in both basins receive about sixty inches of rainfall annually. This is four to five times as much rainfall as the western portion of the country receives. With an abundant amount of annual rainfall, it might seem unlikely that concerns over river water allocation would develop in the Southeast.

Water quality and quantity issues have long been important public policy concerns in the western portion of the United States (Davis 2001; Freeman 2000; Lewis 2006; McCool 1997, 2002; Shurts 2000; Wilkinson 1992). Only recently have water issues caught the attention of policy makers in the South. Given the region's abundance of rivers and underground aquifers, concerns over access to freshwater have been few. As a result, citizen uses of and attitudes toward water reflect a belief that freshwater is abundant and limitless. This abundance of freshwater throughout the history of the region has influenced water law and the evolution of water policy (Dellapenna 2002).

The problem of access to fresh water has its origins at the Chattahoochee River's headwaters in northern Georgia. The river serves as the primary source of drinking water for three million people in the Atlanta metropolitan area. The water source for the remaining two million people living north and west of Atlanta is the Etowah and Coosawatee rivers. Historically, most major cities developed near the mouth of a major river (Jackson 1995; Jordan and Woolf 2006). New York, Baltimore, Washington, DC, and New Orleans are examples of this pattern. To further complicate problems relating to fresh water access, there is no underground aquifer that can be tapped due to the geological constrains of the region.

The city of Atlanta is near the headwaters of the Chattahoochee River and south of Lake Lanier, which serves as a major source of fresh water. The Chattahoochee River also provides drinking water, commercial development, and electrical generation for Alabama residents. The river flows southwest of Atlanta towards Alabama, eventually forming the state line between the two states for over two hundred miles until it combines with the Flint River to form the Apalachicola River in Florida. The Etowah and Coosawatee rivers combine with the Oostanaula east of Rome, Georgia to form the Coosa River which flows into Alabama.

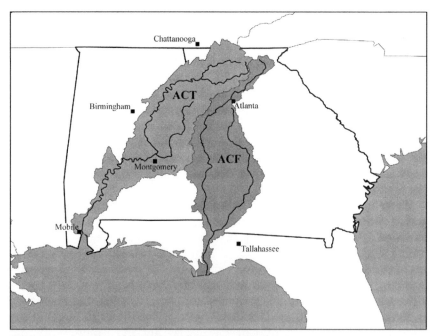

Figure 5.1. Location of the ACT and ACF Watersheds

Cartography by Sarah E. Hinman. Data Sources: ESRI Data & Maps CD 3 2002. United States Department of Agriculture. Geospatial Data Gateway. 8 Digit Watershed Boundary Dataset (HUC 12 subset). Last accessed: 8 February 2010.

This chapter examines the context in which the compacts were created and how the region's views of water rights, and environmental issues as a whole, influenced the direction and scope of the compacts' negotiations. Considering the traditional resistance of federal involvement in environmental policymaking among southern states, using interstate compacts to address environmental policy issues should not come as a surprise. In using interstate compacts to address these concerns, the three states were showing signs of innovation in environmental policy. While interstate compacts have a long history of addressing water issues in the West, compact use in the South is unprecedented as a tool to address environmental policy issues (Dellapenna 2002). While use of interstate compacts in this policy area is not common, what is truly groundbreaking is that the compacts were developed after landmark federal legislation such as the Clean Water Act and the Endangered Species Act. Most existing interstate compacts addressing water allocation were negotiated and approved before the passage of these acts. By inserting federal law into a policy area that had previously been the domain of the states, the negotiations became more complex as legal and political concerns increased (Dellapenna 2002).

Since settlement of western states, disagreements have occurred over access to western rivers and water basins. Many western states have utilized interstate compacts to reach an agreement over legal access to a basin's water. A primary reason for states to use an interstate compact is the hope that it will allow for a decision to be reached without having to expend large amounts of resources on a prolonged legal battle (Lewis 2006).

ELAZAR'S POLITICAL CULTURE

Elazar believes political culture is influenced by the beliefs of early settlers from Europe and can be divided into three primary subcultures: individualistic, moralistic, and traditionalistic. The difference in beliefs among citizens about government's role in society stems from which region of Europe the settlers originated. In short, as the United States was settled by Europeans, they brought their cultural values and political beliefs, which continue to influence contemporary political culture (Elazar 1984). The following is a brief discussion of the three subcultures.

The traditionalistic political subculture originates from an ambivalent attitude toward politics and involves a desire to support a hierarchical society allowing the elites to govern in a paternalistic manner (Elazar 1984). Participants in this subculture trace their background to a preindustrial social order. Typically government plays a custodial and conservative role, not a proactive role. In a traditionalistic culture, citizen involvement in the policy process is often minimal (Elazar 1984).

An individualistic political culture "emphasizes the conception of the democratic order as a marketplace" (Elazar 1984, 230). In this culture, citizens possess utilitarian expectations of government and its interaction with citizens. Citizens residing in an individualistic culture expect government to provide services the public explicitly desires and no more. Because of this view of limited government, primary concerns are not with a greater good. Instead, each individual should be responsible for his welfare (Elazar 1984).

Citizens' expectations of government are similar to their expectations of private business. In an individualistic culture, the process of politics is viewed as a way for citizens to improve their social and economic standing in society (Elazar 1984). Citizens view government's role to be concerned with economic development as a way citizens can improve their lives.

A moralistic culture "emphasizes the commonwealth conception as the basis for democratic government" (Elazar 1984, 232). Citizens in a moralistic culture expect government to promote a public good and treat citizens in an altruistic manner. Participation in government is expected of each citizen. This differs from individualistic and traditionalistic cultures in that participation should be a concern for all citizens, not just individuals who are committed to a political career (Elazar 1984).

To understand the role political culture plays in explaining each state's behavior during the compact negotiations, a review of each state's political culture, migration, and subcultures will be discussed.

Alabama

Elazar (1984) considers most of the state of Alabama to have a traditional culture. While Elazar does indicate pockets of individualistic and moralistic culture exist within the state, traditional political culture is dominant. Since the 1960s when Elazar's data were gathered, Alabama's population growth has been slower than the average rate for the nation (U.S. Bureau of the Census 2003; 1993). This indicates the state will likely be similar in culture as it was forty years ago as the influx of citizens moving into the state is not enough to meaningfully alter the state's culture. In a traditionalistic culture, citizens within the state are likely to have a low awareness of the issues surrounding the compacts, and the negotiation process will go largely unnoticed in state media outlets. Because of the low salience of this issue among citizens, an interest group that is greatly concerned about the negotiations can easily persuade the state to allow it to influence the state's position in the negotiations.

Georgia

Elazar (1984) labels the state of Georgia as a traditionalistic subculture with the Atlanta area as a pocket of individualistic behavior. Georgia's, and in particular Atlanta's, population has greatly increased since Elazar first labeled each state (U.S. Bureau of the Census 2003; 1993). Elazar (1984) believes that as a traditional subculture experiences an influx in migration, it is likely to become more individualistic in nature.

SETTING THE STAGE FOR CONFLICT IN THE NEGOTIATION PROCESS OF THE COMPACTS

The tension between states over the water allocation in the ACF basin can be traced back to the 1980s (Ruhl 2003; Stephenson 2000). Due to a series of droughts in the 1980s along with the continued increase in Atlanta's metropolitan area population, downstream states became concerned about the availability of fresh water (Ruhl 2003; Stephenson 2000). This need for more water prompted several municipalities near the city of Atlanta to ask the Corps of Engineers (COE) for a new reservoir to be located in west Georgia along the Tallapoosa River (Ruhl 2003). Because of concerns about the proposed dam restricting the water flow of the Tallapoosa River, the state of Alabama filed a lawsuit against the COE in hopes of blocking the building of the reservoir. With the future of the reservoir in doubt, municipalities began to increase their demand upon the water in the ACF basin. This soon prompted negotiations of the ACF as well as an ACT compact. A landmark in cooperation occurred in 1997 when Alabama, Florida, and Georgia agreed to an interstate compact that involved allocation of water in the ACF basin which led to the development of the ACT basin interstate compact.

The landscape of the southeastern portion of the United States provides ample access to freshwater rivers and an abundant source of ground water. Unlike the arid western states, states in the Southeast are accustomed to retrieving as much water from rivers as they have desired without concern for the amount of water flowing downstream.

With the exploding growth of southeastern cities such as Atlanta came increasing demands for freshwater. For decades, the Chattahoochee River provided an adequate freshwater supply for metropolitan Atlanta. By the 1990s, the growth of metropolitan Atlanta created a strain on the water level of the Chattahoochee, especially during drought years (Stephenson 2000). The metropolitan area's rapid growth not only increased the amount of water being used, it also increased the amount of pollution being placed into the river.

The metropolitan area surrounding Atlanta has been rapidly growing. This growth has primarily occurred in rural areas with few zoning ordinances. With the growth, a strain on the ability of municipalities to maintain an infrastructure that adequately supplies water to its new residences and industries was created. Many of the longtime residents of the Atlanta area are accustomed to being able to use water in ways that reflect the perceived abundance of this natural resource in the southeastern United States. The residents who have migrated to Atlanta also use water in ways that are indicative of a plentiful resource.

In light of the possibility of lawsuits over water allocation among the three states, many water associations in the southern portion of metropolitan Atlanta began looking to other rivers such as the Flint, which is part of the ACF basin, to retrieve fresh water. The retrieval of water from the ACF basin affected the flow of water into the southern portion of Georgia where the water is used to sustain development and to preserve and enhance agricultural commerce.

The dams and reservoirs along the Coosa and Tallapoosa Rivers, which are located in Alabama, are operated by Alabama Power, which is owned by Southern Company. Southern Company is a holding company consisting of Alabama Power, Georgia Power, Mississippi Power, and Gulf Power (Jackson 1995). Southern Company provides much of the electricity throughout both river basins. If an individual or private business wants water from a reservoir along the Coosa or Tallapoosa Rivers, the entity must ask permission from Alabama Power to withdraw the water. Alabama Power produces more electricity than is demanded from the customers near the dams. Consequently, the surplus electricity is often sold to users in the metropolitan area of Atlanta.

Although the states agreed to create the compacts, they did not agree on details of water allocation within any of the basins. The specifics of the compacts would be negotiated between the states. Once an agreement was reached, the compact would need to be approved by Congress and the president. Throughout the late 1990s and early 2000s, the three states argued over issues central to water policy such as average daily flows, minimum flows of water during periods of drought, and water quality (Jordan and Woolf 2006).

ORIGINATION OF THE COMPACTS

The compacts provided for one chief negotiator and one alternate negotiator from each state and the federal government. Each state was given one commissioner and one alternate commissioner. Although the compacts were distinct and separate, the same commissioners represented the federal government and each state in both compact negotiations and voted on both compacts, with the exception of Florida not having a representative on the ACT compact since none of the ACT basin was within Florida's boundaries. Formally, each state's chief commissioner was its governor. However, each state's alternate commissioner was the person who represented the state at the meetings and voted. By all accounts of each person interviewed, the governor's role was more ceremonial than substantive.

From the outset, the federal government was relegated to the role of an observer. While each state's commissioner had one vote and veto power over the compacts, the federal commissioner did not have a vote. Membership and voting privileges were defined in the original wording of the compacts. One state commissioner explained the role of the federal government as a bystander to the process: "their role was to observe and ensure that federal laws were not being violated; otherwise they didn't speak unless someone had a question" (interview by the author). Each state negotiator had a slightly different explanation for why there was so little federal involvement. However, the common reason given by each of the commissioners was a lack of trust among the states of the federal government in general and the COE in particular.

With the state of Alabama, the lack of trust of the COE arose from the fact that Alabama resented the COE for wanting to construct a dam on a portion of the Tallapoosa River within the state of Georgia without any prior consultation or consent. "Before the compact was created, Alabama felt the COE was running things," said one person familiar with the process of creating the compacts (interview by the author). Not giving the federal government a vote in the compact negotiations appeased Alabama. One state commissioner concurred that Georgia also had some suspicion of the role of the federal government stemming from the concern that the Environmental Protection Agency (EPA) might have reservations over the quantity of a river's water flow once the river left the state (interview by the author).

In summary of the states' positions, each state feared the federal government would override their veto power and side with a different state. This mistrust prevented the federal government from becoming an active partner in the negotiations.

Interest groups from Alabama believed their suspicions of the federal government were justified once the federal commissioner was named. Lindsey Thomas was a former Georgia congressman and ultimately served two years as president of the Georgia Chamber of Commerce during his tenure as the federal commissioner. As one river stakeholder reflected, "Lindsey was a good guy, very bright. But no matter how hard he tried, there was always the appearance of a conflict of interest" (interview by the author).

Once the negotiations began, Georgia hit the ground running. "They came ready to play," remarked one Alabama lake stakeholder (interview by the author). Georgia had the data and the desire to move forward with the negotiations quickly. As one state commissioner explained, "Georgia knew what it needed. Georgia had done the studies and knew how much water it needed to meet future demand" (interview by the author).

This preparedness was in no small part due to the efforts of previous attention given to water allocation and quality issues by local governments in the Atlanta area. Since the 1970s, local governments had studied the capacity of the Chattahoochee River to sustain growth in the Atlanta area. This awareness of water issues within local government was not uniform throughout the basin. While awareness was high in Atlanta, Rome, and Columbus, Alabama cities did not have the same levels of concern. An environmental engineer, who is a planner employed by an Alabama county downstream from the Atlanta area, indicated that while the issues surrounding the compacts were of interest to him, they were not of interest to his employer (interview by the author). This lack of concern by local government in Alabama is consistent with a traditional political subculture.

Alabama was less prepared for the negotiation process than Georgia. It not only did not know how much water it needed, but it also did not have a mechanism for determining its needs. "Alabama was not ready for this in three ways: politically, technically, and legally," commented one commissioner (interview by the author). Alabama did not have an agency in its department of environmental management that addressed water concerns and did not have an agency with any experience in interstate negotiations on environmental issues. It turned to Alabama Power for expertise. Alabama Power operates every dam along the Coosa and Tallapoosa Rivers and was considered by each commissioner and many stakeholders to be a powerful force with great political clout within Alabama's state government (interview by the author). Indeed, the first commissioner for Alabama in the compact negotiations was an attorney who represented Alabama Power in many civil matters. In 1999, when a Democrat replaced a Republican governor, a former Alabama Power hydrologist replaced the state's negotiator. Some stakeholders believed this was outsourcing government authority as well as technical expertise. Again, the tendency of government in Alabama to defer governing to an elite interest group is indicative of a traditionalist political subculture.

Legally speaking, Alabama had no statewide water policy. The only laws addressing water withdrawals required notice for withdrawing more than 100,000 gallons a day. Only a notice of intent to withdraw the water had to be filed. The claim to the water could not be denied. As one person familiar with the negotiations in Georgia commented, "If a state doesn't have the authority to tell its own people to not take water, how can it say 'no' to another state?" (interview by the author). This attitude fueled the belief that Alabama was not prepared for negotiations and was being unreasonable in asking Georgia to do more than Alabama was willing to do in terms of limiting its uses of water (interview by the author).

While Georgia was well prepared to enter the negotiations, the information it had came from Atlanta area needs assessments. From the outset,

Georgia's position became synonymous with what metropolitan Atlanta wanted. "Atlanta fully backs Georgia's position. Our position is their position," remarked one stakeholder from the Atlanta area. Others downriver had a different view (interview by the author). One stakeholder downriver from metropolitan Atlanta echoed Atlanta's influence: "They (Georgia's negotiation team) would take a position and tell us to trust them. They wanted us to believe they had our best interests involved. Actually, Alabama's interests were closer to ours" (interview by the author). This created a political rift that became public. Representatives from municipalities in Georgia down stream from Atlanta, such as Columbus and Rome, would publicly express their dismay about the direction their state was taking. At one point, the city of Columbus expressed a desire to join Alabama in a lawsuit against the COE for allowing Lake Lanier to maintain a water level during drought periods regardless of the water flow downriver (Nix 2004).

"This was really a negotiation among four states: Alabama, Florida, Atlanta, and the rest of Georgia," remarked one Florida stakeholder. Sometimes information given to non-Atlanta stakeholders about the compacts was in scarce supply. One member of Florida's negotiation team commented, "People from Georgia would come to the public meetings down here and wonder why they had to go to Florida to find out what is going on" (interview by the author). This increased the perception if not the reality of mistrust within Georgia of their state's negotiators. Indeed, there were many issues to be worked out within the state of Georgia. At the time, Georgia did not have a state water policy.

After years of negotiation and deadline extensions, talks for the ACF agreement ceased and the compact expired on July 31, 2003. The ACT basin agreement between Alabama and Georgia expired on July 31, 2004 (Jordan and Woolf 2006).

DISCUSSION OF POLITICAL CULTURE
IN ALABAMA AND GEORGIA

The following is a discussion of how Elazar's political subculture explains the process by which each state chose its negotiators and managed the negotiation process. This discussion also considers Alabama and Georgia's desired goals and outcomes for the compacts. The analysis considers data gathered from interviews of individuals intimately familiar with the negotiations and the issues surrounding the compacts.

Florida is not included in the discussion due to a lack of data because of its small population and the small impact of the ACF Basin within the state. The portion of Florida with the Chattahoochee River Basin is sparsely populated compared to the remainder of the state. Also, the portion of Florida within the ACF Basin is small geographically, and the state did not have as large a stake in the negotiations as Alabama and Georgia since the majority of both state's population is within the boundaries of the two basins.

Alabama

The negotiation process that played out in Alabama is consistent with a state possessing a traditional political subculture. In a traditional subculture, personal relationships play a large role in a government's decision-making process (Elazar 1984). Alabama supports this in its relationship with Alabama Power and its ease of transitioning into the role of negotiation. The lack of preparedness on the part of state government indicates it was not particularly concerned about the compact negotiations and their outcomes. Using representatives from a dominant interest group to manage the state's negotiations indicates a willingness to allow policy making to be made by an elite group of citizens with minimal involvement from the population. These actions are indicative of a state with a strong traditionalistic political subculture (Elazar 1984).

When the compacts began in 1997, Republican governor Fob James appointed an employee of Alabama Power to represent the state in its negotiations. This person, James M. Campbell, was retained by Don Siegelman, a Democrat, who was inaugurated in January of 1999. As the governorship changed hands in 2003, with the inauguration of Republican Bob Riley, a different chief negotiator was chosen (Shelton 2003). This new negotiator, Onis "Trey" Glenn III, also had a longtime and significant business relationship with Alabama Power (interview by author).

Finally, decision making in Alabama appeared to be driven by one strong interest group with a narrow focus. This deference of state governance to an interest group with a long-standing relationship with members of both political parties is a clear example of a state with a traditionalistic subculture, in which elites conduct the activities of government. The state's environmental management agency's lack of preparedness before and during the negotiations and subsequent willingness to allow Alabama Power to use its expertise to represent the state indicates support for governing by the elites.

Georgia

Analysis of the interview data indicates strong support that the individualistic political subculture is alive and well in Georgia. Elazar (1984) contends regions in an urban traditionalistic culture that experience growth fueled by migration from other states are likely to adopt an individualistic culture as the traditional subculture breaks down. Furthermore, in regions with a large population of African Americans, characteristics of an individualistic subculture are typical (Elazar 1984). The majority of citizens in Atlanta identify their race as African American (U.S. Bureau of the Census 2003; 1993).

With the large African American population in Atlanta and the influx of new residents throughout Georgia, it is expected that the individualistic subculture dominates local politics and the decision makers in the Atlanta area as well as policy-making decisions in Georgia's state government. Because a traditionalistic subculture evolves into an individualistic subculture as a result of new immigration, it is a natural progression for the Atlanta area to become predominantly individualistic. The evidence from the interview analysis shows a state with a dominant individualistic political subculture.

From the outset, Georgia's interests reflected the economic interests of businesses throughout the metropolitan area of Atlanta with other interests in the state having negligible influence. The state of Georgia was well prepared to begin the negotiations and state government continually took an interest in the process in support of the interests of the Atlanta area through the individuals it appointed to represent the state and its willingness to initiate proposals and take an active role in setting the agenda of the negotiations.

One area of strength for Georgia is that its lead negotiator did not change throughout the life of the compacts. Georgia wanted to have some consistency in its desires and position throughout the negotiation process, and one way to ensure this was to have the same individual represent the state in the compact negotiations. The support from the business interests in the Atlanta area was consistent and strong during the negotiations as the primary goal of protecting the amount of water municipalities could withdraw from the river did not waver. Georgia's goal of providing enough water for the Atlanta metropolitan area for the purpose of continuing economic growth and the predominance of citizens with strong connections to economic development are consistent with an individualistic subculture's values of operating government in a businesslike manner. This approach to the negotiations is consistent with an individualistic culture that takes a business-like approach to governing.

The negotiators for Georgia frequently told other in-state stakeholders to "trust us" (interview by author). Further evidence of placing the economic interests of commercial development in the Atlanta area is illustrated by Georgia supporting a sitting Chamber of Commerce president for the state of Georgia to represent the federal government's interests.

Again, this is consistent with a desire for a business-like approach to managing the government's affairs. Citizens in an individualistic subculture desire to leave the business of government to professionals. In short, politics is not for laypeople. Elazar's individualistic subculture explains the reason Atlanta area business interests were so dominant in representing the state's interests during the negotiations.

The interests of Georgia, which are concerned with protecting economic development issues, likely worked closely with local governments in the Atlanta area as well as the different chamber of commerce organizations. The local governments, which are concerned about promoting their population growth and economic development, are interested in having more access to river water to fuel this growth. These groups, with their vast political resources and similarity of goals, were able to influence Georgia's goals and desired outcomes for the compacts. This agenda was advanced by a narrowly focused, well prepared negotiator.

The state of Alabama chose negotiators who had significant professional connections with Alabama Power Company. Allowing a private company that has a long-standing relationship with the state to represent the state in the negotiation process is indicative of a state with a traditionalistic culture. A state government in a moralistic or individualistic culture would not have been as likely to delegate policy-making authority to a private entity. Because of the negotiators' connections to Alabama Power Company, respondents indicated they were the most influential group in Alabama during the negotiation process.

In summary, government agencies and elected officials at many levels of government in Georgia were more involved in the process and worked closely with business interests in developing Georgia's desired outcomes of the negotiation process. This approach is indicative of a state possessing an individualistic subculture in that government chose to take an active role and openly solicit assistance from other governments and business interests. In an individualistic subculture, government is to take on a business like approach to managing the affairs of government (Elazar 1984). Georgia's focus on economic development for the region and not one particular industry, company, or organization is consistent with a subculture that places great value in economic development and a desire for government to proactively become involved in the process. In this case, Georgia's agenda was getting enough fresh water to promote the continuation of economic development in the Atlanta area. In an individualistic subculture, economic concerns are appropriate spheres of activity for government (Elazar 1984). The approach Georgia took in securing its goals is indicative of a government in a state with an individualistic subculture.

CONCLUSION

While Elazar's (1984) discussion of political culture assists in explaining the behavior of each state during the negotiation process, his theory of political culture best explains the decision-making process of choosing a negotiator and which organizations and interest groups were more influential in the negotiation process. Elazar's theory goes a long way in explaining the different approaches Alabama and Georgia took during the negotiation process and the outcomes each state desired.

Use of interstate compacts to solve policy issues surrounding water quantity and quality indicate that while a resistance to federal involvement continues to be strong, innovation in environmental policy development and implementation is possible at the state level. This theme of uniqueness of innovation in environmental policy may serve as a guide for states as they consider environmental issues in an era of devolution and post landmark environmental legislation such as the Clean Water Act and Endangered Species Act.

The process by which Alabama chose its negotiators and its desired outcomes of the compact's negotiation process are indicative of Elazar's (1984) characteristics of a "traditional" political subculture. The most obvious example of this is the relationships within Alabama state government with Alabama Power. Allowing an interest group that has enjoyed a longstanding relationship with Alabama state government to have a great deal of influence in determining who would represent the state and therefore the agenda the state would pursue during the negotiations is precisely characteristic of Elazar's traditionalistic subculture. Further support of this is the lack of awareness of issues surrounding the compacts among the electorate and state legislators. A characteristic of traditionalistic subcultures include a lack of participation among the populace in politics and the policy-making process. A lack of awareness of this issue among citizens and state legislatures is further evidence of a traditionalistic subculture.

Another supporting characteristic of a traditionalistic subculture is the limited amount of knowledge of the negotiations made available to the public and municipalities that receive drinking water from the Coosa River from the negotiators. This in turn caused a low amount of media coverage of the negotiation process. In part because of low media attention, there was a lack of awareness of the issues and few people knew the specific issues being discussed. Because of the lack of awareness and knowledge of the issue, there was little public involvement in the process. This lack of interest and involvement in the policy process by citizens is consistent in keeping with what Elazar (1984) labels as a traditionalistic subculture.

Evidence of Georgia's individualistic subculture is seen in the state taking a business-like approach in choosing the state's negotiator and the active relationship of state and local government with the state's negotiator. In Elazar's original collection of data, the state of Georgia was described as a traditionalistic society with individualistic pockets in the Atlanta area (Elazar 1984). As migration to an area increases, Elazar (1984) contends the area will become more individualistic. Given Georgia's rapid increase in population around the Atlanta area, using Elazar's theory, it is expected that Georgia will display characteristics that are consistent with an individualistic subculture.

Georgia approached the negotiations better prepared than Alabama and with narrowly focused goals. This organized approach with a focus on economic development issues within the Atlanta area is reflective of an individualistic subculture. As the Atlanta area has experienced a great migration of citizens, the individualistic subculture has increased its influence over the area's government and consequently is quite influential in Georgia state government. While it is obvious Georgia's negotiation goals were driven by economic development interests from the Atlanta area, the approach toward the negotiations in choosing a commissioner and the proactive relationship between government and economic development interests is indicative of the state having followed Atlanta's evolution from a traditionalistic to an individualistic subculture.

Media coverage of the negotiations within the state of Georgia was greater than in Alabama. Media outlets in different cities from Atlanta to Columbus along the Chattahoochee River frequently discussed the negotiation process. This media attention increased citizen awareness and ultimately involvement. This level of involvement in the process illustrates a clear move away from a traditionalistic subculture.

This study of Georgia's decision-making process during the compact negotiations illustrates the evolution from a traditionalistic to an individualistic subculture. This is consistent with Elazar's (1984) theory of the evolution of the political culture of a state with a traditionalistic subculture with a migration of individuals from other states into an urban area. The findings of this research support Elazar's theory of political subcultures. Alabama and Georgia displayed many characteristics of their respective subcultures as described by Elazar.

EPILOGUE

On July 17, 2009, Senior United States District Judge Paul Magnuson determined municipalities surrounding Atlanta did not have a right to continue withdrawing water at current levels (Rankin 2009). The judge ruled Buford Dam was not created to supply water for much of metropolitan Atlanta's current needs. The dam's purpose was to provide electricity, flood control, and navigation assistance. Judge Magnuson gave state governments in Alabama, Georgia, and Florida or the United States Congress three years from the date of his ruling to reach an agreement for water sharing. If no agreement is reached, withdrawal levels for metropolitan Atlanta will return to levels not seen since the 1970s when the river supplied water to almost half the number of current users. Currently approximately three and a half million people receive drinking water from the Chattahoochee River (Redmon 2010). Judge Magnuson further iterated only water authorities in Gainesville and Buford, Georgia, are authorized to withdraw water from Lake Lanier to meet current and future water needs.

The fate of water sharing throughout the two basins is in the hands of the courts. On January 21, 2010, a federal court rejected a request on the behalf of the states of Alabama and Florida that an appeal of the ruling be denied. This decision from a federal court gave life to legal battles among the three states despite efforts made by the three governors to reach a settlement out of court. One notion is certain: Arguments over water allocation in the Deep South are not going to be solved any time soon.

REFERENCES

Davis, S. 2001. The Politics of Water Scarcity in the Western States. *The Social Science Journal* 38, 4 (Winter):527–42.

Dellapenna, J. 2002. The Law of Water Allocation in the Southeastern States at the Opening of the Twenty-First Century. *University of Little Lock Law Review* 25, 9:9–88.

Elazar, D. J. 1984. *The American Mosaic*. Boulder, CO: Westview Press.

Freeman, D. 2000. Wicked Water Problems: Sociology and Local Water Organizations in Addressing Water Resources Policy. *Journal of the American Water Resources Association* 36, 3 (June):483–91.

Hart, A. 2003. Water Talks End without Deal. *New York Times*, 2 September 2003, 16.

Hyatt, R. 2004. Tug-of-Water: Fighting over the Chattahoochee. *Columbus Ledger-Enquirer*, March 21, 2004, 3.

Jackson, H. 1995. *Rivers of History. Life on the Coosa, Tallapoosa, Cahaba, and Alabama*. Tuscaloosa: University of Alabama Press.

Jordon, J. L., and A. T. Woolf. 2006. *Interstate Water Allocation in Alabama, Florida, and Georgia: New Issues, New Methods, New Models*. Gainesville: University Press of Florida.

Lewis, L. 2006. Interstate River Compacts of the West. In *Interstate Water Allocation in Alabama, Florida, and Georgia: New Issues, Methods, New Models*, edited by J. L. Jordan and A. T. Wolf, 102–30. Gainesville, FL: University Press of Florida.

McCool, D. 1997. *Command of the Waters. Iron Triangles, Federal Waters Development, and Indian Water*. Berkley: University of California Press.

———. 2002. *Native Waters*. Tucson: University of Arizona Press.

Nix, D. 2004. Columbus Splits with State over Tri-state Water Deal. *Columbus Ledger-Enquirer*, February 6, 2004, 1.

Rankin, B. Judge Again Rules against Georgia in Water Fight. *Atlanta Journal-Constitution*, www.ajc.com/news/federal-judge-rules-against-94051.html (accessed on October 6, 2009).

Redmon, J. Judge: State's Water Talks Can Be Secret, *Atlanta Journal-Constitution,* www
.ajc.com/news/judge-states-water-talks-270323.html (accessed on January 9, 2010).

Ruhl, J. B. 2003. Equitable Apportionment of Ecosystem Services: New Water Law for
a New Water Age. *Journal of Land Use and Environmental Law,* 19, 1 (Fall):47–57.

Shelton, S. 2003. Water Wars: GOP Shift New Reality as Tri-state Talks Begin. *Atlanta
Journal-Constitution,* January 6, 2003.

Shurts, J. 2000. *Indian Reserved Water Rights, The Winters Doctrine in Its Social and
Legal Context, 1880s–1930s.* Norman: University of Oklahoma Press.

Stephenson, D. 2000. The Tri-State Compact: Falling Waters and Fading Opportunities. *Journal of Land Use and Environmental Law,* 16, 1 (Fall):83–109.

U.S. Bureau of the Census. 1993. *Census 1990.* Washington, DC: Author.

———. 2003. *Census 2000.* Washington, DC: Author.

Wilkinson, C. 1992. *Crossing the Next Meridian.* Washington, DC: Island Press.

6

Collaboration in Environmental Policy Implementation

Brownfields Programs in North Carolina and Florida

Deborah R. Gallagher

INTRODUCTION

Since the 1980s public services traditionally provided by local, state, and federal governments have regularly been contracted out to private entities (Kodras 1997; Kettl 1993). For example, in many cities waste management is no longer the domain of public works departments, but of private contractors (Miranda and Andersen 1994). Federal environmental and natural resource management agencies employ contractors to study contamination at Superfund sites (Kettl 1993) and the behavior of salmon in national parks (Tremain 2003). These activities are typically governed by detailed contracts. Increasingly, however, businesses are working with public sector agencies to interpret and implement complex public policies, with regulations and guidance in hand, but often without the benefit of explicit contracts. The relationship between business and government in these cases is best described as a partnership or collaboration. In the American South, where tradition and innovation exist side by side (Black 1987, 297), such collaborations are both steadfast and precarious as partners seek to protect traditionally close business-government relations, while reaching out to the broader community to develop innovative solutions to vexing problems.

Research on public/private partnerships has focused on creating broad frameworks for understanding the phenomenon and suggesting criteria to be used in evaluating partnership success. Scholars have described specific types of public-private policy partnerships and pointed out their strengths and weaknesses. Rosenau (1999) suggests that while public/private partnerships may provide short-term cost-savings, they may not adequately address equity and democratic participation issues. Linder

(1999) suggests that such partnerships fall into six distinct categories: management reform, problem conversion, moral regeneration, risk shifting, restructuring public service, and power sharing. For example, under the guise of management reform, firms assist public sector partners in establishing market mechanisms to solve public problems like air pollution. In a problem conversion setting, public sector managers become entrepreneurs by redefining public sector problems as opportunities for profit. Incentive structures are created to lure private sector collaborators into underperforming markets. In brownfields programs private sector entrepreneurs are provided regulatory and financial incentives to address the problem of blighted and underused real estate.

The challenges of collaboratively implementing public policies are many. They include development, implementation, and oversight of productive public sector–private sector relationships. An additional challenge is the development of mechanisms to evaluate public outcomes like equal access to products and services. Challenges include the configuration of relationships between the actors, the ability of the public sector actors who recruit private sector partners to oversee the relationship, and the need to engage an objective entity, such as an expert commission, to be engaged in oversight. Additionally, mechanisms to evaluate outcomes must be developed and procedures for public participation in decision making must be established.

Private sector partners, recruited for their abilities to "get the job done," are not always experts at evaluating public outcomes, accustomed to providing transparency, or skilled in providing access to those outside of their internal organizational decision-making processes. However, public projects require that attention be paid to such activities. When responsibility is devolved from public sector organizations to private sector partners, those partners must implement processes to address established public values.

This paper considers the setting of abandoned hazardous waste sites, where private sector organizations have long been involved in assessing contamination and constructing remedial clean up technologies. A recent phenomenon in this domain is the further devolution of responsibility to private sector partners, who seek financial benefits from full scale assessment, remediation, and redevelopment of such sites.

This paper examines how two southern states, North Carolina and Florida, have designed and implemented brownfields partnerships. It considers how the two states' brownfields redevelopment programs address environmental justice objectives while preserving the traditional primacy of economic development. First, background on brownfields redevelopment is provided, followed by an analysis of critical issues related to public/private sector collaboration in brownfields redevelopment. Next, each state's brownfields program is described. The programs are then compared and contrasted from a comparative institutional analysis perspective. Finally, lessons from these programs and recommendations for public policy are offered.

PUBLIC-PRIVATE PARTNERSHIPS FOR BROWNFIELDS REDEVELOPMENT

Brownfields programs are the latest phase in a movement to privatize waste site remediation activities, which began with the federally operated Superfund program in 1980. Superfund incorporated mechanisms to order site owners to take the lead in remediation, but by the mid-1990s citizens and policy makers were increasingly dissatisfied with Superfund's progress (Simons 1998). In response, voluntary cleanup programs, or VCPs, were developed by many state governments. These programs call on site owners to voluntarily remediate contamination in exchange for limited state oversight. Brownfields programs, extensions of state VCPs which seek to further privatize and simplify remediation efforts, were initiated soon after. Brownfields programs provide regulatory and tax incentives, liability reduction, and real estate development opportunities to purchasers who seek to clean up parcels of contaminated, unused, or underutilized urban real estate for a profit. The federal Small Business Liability Relief and Brownfield Revitalization Act (H.R. 2869) was enacted in 2002 to provide limited liability and tax relief to prospective developers, and most U.S. states have developed programs to encourage businesses to purchase, clean up, and redevelop contaminated sites. Participant firms enter into agreements with state agencies and work with government officials to remediate properties in accordance with local zoning and state and federal environmental requirements appropriate for the redeveloped site's end use.

It has been estimated that perhaps as many as 500,000 brownfield properties exist in the United States (Lee and Seago 2002). Most states have implemented brownfields programs that include authorizing legislation and dedicated staff (USEPA 2005). State programs range from California's 1997 initiative, which estimates that it will address over 90,000 brownfields sites, to South Dakota's federal grant supported program, begun in 2002 to focus on three sites. Most state programs offer potential developers attractive incentive programs, such as protection from future liability once the site is remediated to future use status, simplified site cleanup procedures and standards, limited government oversight, and in some cases, loans from state-supported brownfield redevelopment funds. Some states, concerned about preserving vehicles for ensuring public participation and promoting environmental equity established in the Superfund program, offer technical assistance to communities working with brownfields developers.

State program personnel act primarily as facilitators in the context of brownfields initiatives. In this setting, the problem of remediating abandoned hazardous waste sites is converted from one of public crisis to private sector opportunity. State agency staff members become entrepreneurs who focus on education and outreach to potential private sector investors. Once priority parcels for redevelopment are identified and investors come forward, collaboration between private sector developers and state agency personnel focuses on certifying eligibility for public support for matters such as liability limitation and loan payments, negotiating brownfield cleanup agreements, and overseeing remediation progress. The investors themselves take primary responsibility for working with the host community to ensure that local zoning procedures such as public participation are followed as the site is redeveloped.

Frameworks for interdependence set the tone for collaborative relationships between public sector environmental agency or economic development staff and private sector real estate investors addressing brownfield properties. A well-developed and implemented framework that fosters trust and creates accountability between partners is necessary in the brownfield context. This is because private sector partners provide primary leadership in addressing the brownfield dilemma, while public sector managers serve as reviewers of private sector activities.

Collaboration frameworks often come in the form of "brownfield agreements," which set out the responsibilities of each party as established by state regulation and guidance. Because they are legally binding documents they also include milestones and standards which enable public sector partners to measure progress and monitor the effectiveness and public acceptability of outcomes. Such standards ensure the sustainability of the collaboration in the long term. This is especially important because brownfield remediation efforts typically take place over a period of at least five years, and the long-term success of the collaborative outcome ultimately depends on public acceptance. To further complicate matters, brownfields sites are often located in socioeconomically disadvantaged communities, where issues of environmental justice and equity have long been a challenge (Bullard 1994). Private sector partners working in these communities must engage in activities to both mitigate past environmental harm and engender support for remediation and redevelopment outcomes from an often skeptical public.

Florida and North Carolina's brownfields programs are analyzed in the following section. While, admittedly, many variables could be considered in evaluating the effectiveness of private implementation of public policies for hazardous waste site remediation and redevelopment, two issues critical to evaluating the effectiveness of public-private partnerships in socioeconomically disadvantaged communities are emphasized. First, the creation and maintenance of collaborative state agency–private sector relationships are considered. Second, the design and implementation of opportunities for public participation which foster considerations of environmental justice and equity are examined.

FLORIDA'S BROWNFIELDS REDEVELOPMENT PROGRAM

Florida's Brownfield Redevelopment Program emphasizes simplified procedures for site remediation and incentives such as tax breaks and low and no interest loans to private sector businesses willing to undertake remediation. The Florida Brownfield Act was passed by the state legislature in 1997 for the purpose of "reduction of public health and environmental hazards from abandoned and underused commercial and industrial sites, creation of incentives to encourage voluntary cleanup and redevelopment design and deployment of risk-based cleanup procedures and creation of opportunities for environmental equity and justice." The Florida Department of Environmental Protection administers the program. Since the brownfield program's inception, local governments in Florida have designated 142 areas as having potential for brownfield redevelopment. These areas range from small, one-acre plots in rural counties to 10,000+ acre sites encompassing large portions of cities. Designated areas do not represent acreage in which clean up is required, but rather acreage eligible for regulatory and economic incentives under the Act, as described below.

Brownfield site designation, the first step in the redevelopment process, establishes relationships between state agency managers and parties responsible for remediation. Designation is the most critical component of the program and may be initiated either by local governments or individuals, termed Persons Responsible for Brownfield Site Rehabilitation. Individuals initiating designations are generally owners of underused, contaminated land or entrepreneurs interested in redeveloping it. Once a site is designated, financial and regulatory benefits are generally available, but in most cases a more formal Brownfield Site Rehabilitation Agreement (BSRA) is required for specific benefits to be accessed. The agreement, negotiated between Florida's Department of Environmental Protection and the parties responsible for redevelopment, addresses liability issues and administrative duties of local governments related to site clean up including collaboration with department officials, technical components of remediation efforts, including contractor qualifications and procedures for certifying the site as complete, establishment of an advisory committee, and a schedule for completion.

Both regulatory and financial incentives are available to program participants. For example, they may obtain federal incentives through the U.S.

Department of Housing and Urban Development, which offers Brownfield Economic Development Initiative funds and loan guarantees through the Community Development Block Grants program. Additionally, Clean Water State Revolving Loan funds are capitalized with federal monies and provide low interest or no interest loans to program participants. State incentives include liability protection for lenders and redevelopers who execute a BSRA. A Voluntary Cleanup Tax Credit provides tax credit in tangible personal property or corporate income with the execution of a brownfields agreement. A $2,500 job creation bonus refund for each job created in a brownfield area by an eligible business is also offered. The Florida Department of Revenue also offers a sales tax credit on building materials used for construction of a redevelopment project located in an urban high crime area, enterprise or empowerment zone, front porch community, designated brownfield, or urban infill area. Finally, local incentives for brownfield redevelopment may be provided.

Regulatory incentives include simplified rules and streamlined procedures for addressing site contamination. In 2006 the Florida Department of Environmental Protection issued revised guidance which simplified clean up rules and standards for brownfield remediation. These standards resulted from the work of a legislative workgroup which considered the use of risk-based corrective action cleanup standards as an alternative to established compound specific clean up target levels. Private sector brownfield program partners had considered the clean up targets to be overly restrictive and preferred a risk-based approach, which offered opportunities for making site-specific cleanup judgments based on relative risks and reducing remediation costs.

NORTH CAROLINA'S BROWNFIELDS REDEVELOPMENT PROGRAM

North Carolina's Brownfield program is a market-based approach to hazardous waste site clean up targeted at potential developers of abandoned, contaminated, and undervalued real estate. The program provides tax incentives and liability protection for brownfield redevelopers. The program's authorizing legislation, the Brownfields Property Reuse Act, was enacted in 1997 to encourage redevelopment. Benefits under the Act are not extended to owners of uncontrolled hazardous waste sites; the Act explicitly states that there must exist a "public benefit commensurate with the liability protection provided" to developers.

North Carolina's program is dependent on prospective developers initiating partnerships with the Department of Environment and Natural Resources. The state has not identified potential sites, but depends directly on developers contacting the Department to locate sites. The Department has developed a marketing brochure and holds conferences and seminars to attract potential developers to the program. During the first years of the program each prospective developer wishing to enter into a brownfield partnership with the state was required to submit a letter of intent to apply for the program. The letter described proposed land use, remedial actions, and land-use restrictions, if applicable. Following the letter, accepted developers negotiated site-specific brownfields agreements with the state.

The letter of intent procedure no longer exists; however, the Department simplified its brownfields procedures to encourage increased participation. Developers now submit brownfields property application forms which provide sufficient information for the Department staff to determine whether the developers' proposed projects would be eligible to enter into a negotiation to establish a site specific brownfields agreement with the state. Once an agreement is reached developers must adhere to its specific terms and the requirements of any prior site remediation, as well as federal and state environmental and public health rules and regulations and procedures. The developer must show sufficient financial capacity to complete the project and must demonstrate that public benefits from the project are commensurate with the liability protection it is seeking. Public participation tools are not often employed to gauge whether community members support proposed projects and in fact public benefit is often assumed. State program managers believe that any remediation is in and of itself a public benefit, stating that in reviewing the application they "know it when they see it."[1]

In addition to the federal incentives such as Brownfield Economic Development Initiative funds and Community Development Block Grant loan guarantees, North Carolina offers two additional financial incentives and a procedural incentive to prospective developers. A covenant not to sue, whereby prospective developers will not be held liable "for remediation of areas of contaminants identified in the brownfields agreement except as specified in the brownfields agreement, so long as the activites conducted on the brownfields property by or under the control or direction of the prospective developer do not increase the risk or harm to public health or the environment," is provided under the brownfields agreement. In addition, tax reductions are provided to developers in the form of reduced property taxes for a period of five years.

Procedural incentives center on a simplified process for site remediation managed by a dedicated brownfields program staff. Developers pay fees to the department for carrying out the brownfield project review procedure, which is designed to minimize lengthy and detailed oversight of developer actions and to clearly delineate roles and responsibilities of both partners in the collaboration. These fees support the hiring of sufficient technical staff to manage the program at an optimal level.

COLLABORATION, PARTICIPATION, AND ENVIRONMENTAL EQUITY

Collaboration is a process of making joint decisions about what future actions to take and how those actions are to be implemented (Gray 1989). Public/private sector collaboration for social problem solving is dependent on the use of existing communication networks and on the development of new networks between partners and external stakeholders to generate new ideas (Waddock 1989; Hood et al. 1993). When public and private sector organizations form partnerships to collaborate on social problems such as brownfield redevelopment, institutional arrangements must be designed to engender communication between the public/private partners and between the partners and community stakeholders.

The success of public private partnerships for brownfield redevelopment, like other types of collaboration, is predicated on a relationship of interdependence between partners and the legitimacy and stability of joint interests (Oliver 1990). These characteristics ensure the long-term sustainability of the partnership. Some researchers have suggested that without firmly established regulation to guide public/private partnerships, they may fail due to information asymmetry and increased complexity (Pongsiri 2002). However, others have suggested that when partnerships are based on trust and common goals, formal arrangements are not always necessary to prescribe roles and responsibilities (Hood et al. 1993).

State level brownfield redevelopment partnerships are alternatives to the predecessor Superfund program, which is overly prescriptive and regulation driven. Superfund's myriad rules, regulations, and guidance documents precisely dictate roles and responsibilities of partners and lay out detailed procedures and technical specifications for site remediation. In contrast, site-specific, jointly negotiated agreements, which frame the roles and responsibilities of each partner and provide flexibility in developing remediation and redevelopment solutions, are the cornerstones of brownfield programs. The agreements are not dictated by detailed regulation, but are generally developed from a template provided by the public sector partner. Templates may include documentation of past remediation, efforts, legal agreements with state agencies, certification of contractors and laboratories, site redevelopment plans, and requirements for public notice or public participation.

Brownfield redevelopment programs address the problem of blighted and contaminated properties. While these sites may be located in rural or suburban areas, many are located in depressed urban neighborhoods (Greenberg et al. 2000), primarily populated by minority and low-income residents (Eisen 1996). Brownfield redevelopment is therefore acknowledged as a public policy tool to address environmental justice concerns whereby disadvantaged citizens suffer disproportionately from the ravages of environmental contamination (Freeland 2004; Byrne and Scattone 1999; Davis 1999). In cases where brownfield redevelopment is employed to address environmental justice concerns, an additional level of collaboration is necessarily required. Public/private partners must work with neighbors to ensure that the site is safely remediated and that redevelopment plans incorporate land uses which are acceptable to neighbors and promote environmental justice goals such as the creation of opportunities for economic development.

This second collaboration presents a challenge to state agency/private entrepreneur partnerships. State-managed brownfield programs may not have sufficient institutional or financial resources to implement a robust public participation program, such as those incorporated in the federally managed National Brownfield Pilot Projects or the Superfund program. These programs incorporate specific rules and procedures for participation, training, and outreach education budgets, and in some cases provide grants to neighborhood groups engaging with remediation and redevelopment staff. Within these programs, participation components are required in order for site remediation to proceed. While state agencies overseeing brownfields redevelopment programs recognize the importance of stakeholder outreach and meaningful public participation, resource constraints often dictate that responsibility for developing a meaningful collaboration with neighbors is left to private entrepreneurs and local governments.

FLORIDA AND NORTH CAROLINA
BROWNFIELDS PROGRAMS COMPARED

Comparative institutional analysis (Aoki 2001) applies a systematic framework from which to compare economic activities such as markets, contracts, or labor relationships. Such a framework considers key components of institutions, from formal rules to implicit behavioral norms, to understand how forms of governance may diverge, even when confronting similar challenges. The following analysis considers the nature of collaborations between agency staff and private sector actors in brownfields programs in North Carolina and Florida. Both formal rules and informal relationship behaviors are examined to understand how each program evolved to become a unique governmental institution. Specifically, the manner in which relationships between private sector firms and public sector organizations are conducted and the manner in which relationships between private sector businesses and the public at large are conducted are compared and contrasted.

Relationships between Private Sector Firms and Public Sector Organizations

Florida's business-government brownfields collaborations are formally structured to promote environmental cleanup, economic development, and environmental justice goals. Partnerships between Florida agency staff and private sector brownfield entrepreneurs are circumscribed by the demands of site-specific, individually negotiated Brownfield Redevelopment Site Agreements. A review of fifty-four Agreements executed between 1997 and 2003 shows that while they have been constructed from a single standard template, they are also designed to address site-specific concerns. The concerns highlighted in the Agreements include issues such as prior remedial actions and existing consent decrees, detailed local government roles and responsibilities, and schedules for document submittals. The concerns that are emphasized in each Agreement set the agenda for the public agency/brownfield entrepreneur partnership. The cleanup phase of the redevelopment project proceeds in a standard four-to-five phase process encompassing site assessment to closure. State agency personnel review progress throughout each phase and are responsible for providing comments at twenty-six potential review points according to well-defined turnaround times. By 2009, 131 Agreements had been executed in Florida, and thirty-one were completed.[2]

In North Carolina, collaborations between state agency personnel and redevelopment entrepreneurs are more flexible and less formal than those in Florida. This flexibility is provided to increase opportunities to enhance economic development. North Carolina's Brownfield Agreements are based on a model framework that addresses concerns such as prior remedial actions and identifies high priority issues, such as land-use restrictions. A simple process model dictates when and how interaction between the public and private partners will occur and cleanup proceeds according to a risk-based framework. North Carolina's program provides just five discrete opportunities for agency review of redeveloper actions. In large part due to this simpler review system, in North Carolina by 2009 there were 109 active brownfields projects, 33 projects pending eligibility, and 109 finalized Brownfield Agreements.[3]

Relationships between Private Sector Firms and the Public

In addition to differences in the nature of public/private partner collaborations within the Florida and North Carolina programs, the ways in which the state agencies manage and direct partner relationships with the public are significantly distinct. When an individual participating in Florida's Brownfield Redevelopment Program initiates an Agreement, notice and opportunity to comment on cleanup criteria must be provided to neighbors and nearby residents. Alternately, if the Agreement is initiated by a local government, one public hearing must be held. Each Agreement designates an advisory committee comprised of residents and businesses adjacent to or within the proposed brownfield area. The committee reviews and makes recommendations on the proposed Agreement and remains active throughout the remediation and redevelopment process. A review of the makeup of advisory committees convened to date shows that many are comprised of business and local government representatives, although some include unaffiliated neighbors and representatives of faith-based organizations. Some of Florida's public participation efforts are far reaching. For example, the city of Clearwater's program, which evolved from its participation as one of EPA's National Brownfield Pilot Projects, has been praised for its efforts to incorporate specific provisions to address environmental justice issues, such as enhanced information sharing and training (NAPA 2002).

In contrast to Florida, North Carolina's program is a more traditionally southern one, with its primary focus on economic development rather than on community involvement. North Carolina's public participation process is founded on a series of public notices. Brownfield entrepreneurs are responsible for providing public notice of their intent to develop in the state register and in local newspapers. A public comment period follows, after which the state agency partner then determines if a public hearing is needed. Few comments are received and public hearings are not commonly held.

While Florida's program goes far to ensure that public participation opportunities are provided to neighbors of remediation sites, neither state program has incorporated formal procedures to fully meet the many challenges that ensue when a public participation program seeks to address environmental justice concerns in socioeconomically disadvantaged neighborhoods. In such settings, residents may lack knowledge of traditional public participation procedures and may not be fluent in English. Researchers have shown that a level of trust between brownfields redevelopment partners and neighbors in poor communities does not necessarily exist and that extra care must be taken to build bridges and provide services such as translators and process education (Solitare 2005; Greenberg and Lewis 2000). However, when partners work closely with neighbors to encourage and support meaningful dialogue, benefits accrue to both parties (Gallagher and Jackson 2008). Remediation and redevelopment efforts proceed more smoothly and outcomes are more likely to be acceptable to neighbors.

Recommendations for Public Policy

This comparison of the Florida and North Carolina state brownfield programs highlights variability between public/private partnerships designed to address brownfield redevelopment in two southern states. It offers a glimpse into how two competing sets of southern political values, tradition, and innovation are incorporated in the design of progressive policies for environmental protection. Entrepreneurs in Florida and North Carolina are offered two distinct models for public/private collaboration. North Carolina's more traditional program is focused on economic development; its highly flexible program incorporates a small number of prescriptive requirements and few public participation obligations. North Carolina provides little opportunity for public participation, unless individual developers see fit to incorporate it or individual communities' planning and zoning processes require it. In contrast, Florida offers a more tightly designed program with a lengthy designation process, more robust procedural and technical remediation requirements and the establishment of a site advisory committee. Given its dual commitment to traditional economic development and innovative community involvement opportunities, it is easy to understand how Florida's brownfield project completion rate is less than that of North Carolina.

These contrasts raise important public policy questions. First, if state brownfield programs are to supplant the federal Superfund program and be used as a primary policy tool for waste site remediation, are flexible public/private sector collaborations based on brownfield agreements sufficient to ensure both the long-term viability of collaboration and the social and environmental sustainability of the redevelopment outcome? In the southern setting, this translates to a question of whether innovation and tradition can both play a leading role in brownfields partnerships. Second, if environmental justice issues are to be addressed by an innovative brownfields redevelopment framework in which public policies are implemented through public/private partnerships, is it appropriate for public participation requirements to be ad hoc, as in North Carolina, or must they be prescribed, as in Florida? Should enactment of relationships between neighbors and developers be wholly devolved to local authorities, or is a more centralized approach appropriate? And finally, does the inclusion of an initial designation process, in which local authorities and entrepreneurs contribute to decisions about redevelopment priorities, assist in ensuring that collaborations and outcomes are sustainable? The answers to these questions are critically important to ensure that today's brownfield programs do not generate a new generation of future abandoned or underused properties. Effective collaboration will go far to prevent that outcome.

NOTES

1. Private communication with Bruce Nicholson, North Carolina Brownfields Program director.

2. Florida Department of Environmental Protection Brownfield Redevelopment Program, http://tlhwww4.dep.state.fl.us/waste/categories/brownfields/pages/monthly.htm (accessed December 2009).

3. North Carolina Department of Environment and Brownfields Program, www.ncbrownfields.org/project_inventory.asp (accessed December 2009).

REFERENCES

Aoki, M. 2001. *Toward a Comparative Institutional Analysis*. Cambridge, MA: MIT Press.

Black, E. 1987. *Politics and Society in the South*. Cambridge, MA: Harvard University Press.

Bullard, R. D. 1994. *Dumping in Dixie: Race, Class, and Environmental Quality*. Boulder, CO: Westview.

Byrne, J., and R. Scattone, R. 1999. *The Brownfields Challenge: Environmental Justice and Community Participation Lessons Learned from National Brownfields Pilot Projects*. Center for Energy and Environmental Policy (CEEP), University of Delaware.

Davis, L. 1999. Working toward a Common Goal? Three Case Studies of Brownfields Redevelopment in Environmental Justice Communities. *Stanford Environmental Law Journal* 18:285–329.

Eisen, J. B. 1996. Brownfields of Dreams? Challenges and Limits of Voluntary Cleanup Programs and Incentives. *University of Illinois Law Review* (1996):883–982.

Freeland, W. T. D. 2004. Environmental Justice and the Brownfields Act of 2001: Brownfields of Dreams or a Nightmare in the Making? *The Journal of Gender, Race and Justice* 8 (Spring):183–214.

Gallagher, D. R. and S. Jackson. 2008. Promoting Community Involvement at Brownfields Sites in Socio-Economically Disadvantaged Neighborhoods. *Journal of Environmental Planning and Management* 51, 5 (September):615–30.

Gray, B. 1989. *Collaborating: Finding Common Ground for Multiparty Problems*. San Francisco: Jossey-Bass.

Greenberg, M., K. Lowrie, L. Solitare, and L. Duncan. 2000. Brownfields, TOADS and the Struggle for Neighborhood Redevelopment: A Case Study of the State of New Jersey. *Urban Affairs Review* 35, 5 (May):717–33.

Greenberg, M., and M. J. Lewis. 2000. Brownfields Redevelopment, Preferences and Public Involvement: A Case Study of an Ethnically Mixed Neighborhood. *Urban Studies* 37, 13 (December):2501–14.

Hood, J., J. Logsdon, and J. K. Thompson. 1993. Collaborations for Social Problem Solving: A Process Model. *Business and Society* 32, 1 (March):1–17.

Kettl, D. F. 1993. *Sharing Power: Public Governance and Private Markets*. Washington, DC: Brookings.

Kodras, J. 1997. Restructuring the State: Devolution, Privatization and the Geographic Redistribution of Power and Capacity in Governance. In *State Devolution in America: Implications for a Diverse Society: Urban Affairs Annual Review*, edited by L. Staeheli, J. Kodra, J. Flint, and C. Flint. 79–96. Thousand Oaks: Sage.

Lee, J. E., and W. E. Seago. 2002. Policy Entrepreneurship, Public Choice and Symbolic Reform Analysis of Section 198, The Brownfields Tax Incentive, Carrot or

Stick or Just Nevermind. *William and Mary Environmental Law and Policy Review* 26:613–83.

Linder, S. 1999. Coming to Terms with the Public-Private Partnership: A Grammar of Multiple Meanings. *American Behavioral Scientist* 43, 1 (September):35–53.

Miranda, R., and K. Andersen. 1994. Alternative Service Delivery in Local Government, 1982-1992. In International City/County Management Association (ICMA). *The Municipal Year Book, 1994.* Washington, DC: ICMA.

National Academy of Public Administration. 2002. *Models for Change: Efforts by Four States to Address Environmental Justice.* Working paper. Washington, DC: Author.

Oliver, C. 1990. Determinants of Interorganizational Relationships: Integration and Future Direction. *Academy of Management Review* 15, 2 (April):241–65.

Pongsiri, N. 2002. Regulation and Public-Private Partnerships. *The International Journal of Public Sector Management* 15, 6:487–95.

Rosenau, P. V. 1999. The Strengths and Weaknesses of Public-Private Policy Partnerships. *American Behavioral Scientist* 43, 1 (September):10–34.

Simons, R. A. 1998. *Turning Brownfields into Greenbacks.* Washington, DC: Urban Land Institute.

Solitare, L. 2005. Prerequisite Conditions for Meaningful Participation in Brownfields Redevelopment. *Journal of Planning and Management* 48, 6 (November):917–35.

Tremain, K. 2003. Pink Slips in the Parks: The Bush Administration Privatizes Our Public Treasures. *Sierra* 88, 5 (September–October):26–33.

U.S. Environmental Protection Agency. 2005. *State Brownfields and Voluntary Response Programs: An Update from the States.* Washington, DC: USEPA (publication Number: EPA-560-R-05-001).

Waddock, S. 1989. Understanding Social Partnerships: An Evolutionary Model of Partnership Organizations. *Administration and Society* 21, 1 (May):78–100.

7

Agricultural Workers and Environmental Justice

An Assessment of the Federal Worker Protection Standards

Celeste Murphy-Greene

INTRODUCTION

The environmental movement has come a long way since it began in the 1960s. One environmental issue that gained national attention in the 1990s is environmental justice. Environmental justice combines both social justice and environmental concerns into one. The environmental justice movement has southern roots, beginning in 1982 in Warren County, North Carolina, with the protest of a toxic landfill. Today, the movement has gained national and international recognition. This chapter will explore the issue of environmental justice today in the South—where it all began—and will discuss how the South still has one of the worst records in regards to environmental justice as it pertains to the Federal Worker Protection Standards. The chapter will begin with a chronological history of the environmental justice movement from its beginning to the present. Next, the issue of environmental risk assessment will be discussed. The chapter will then focus on the case of Florida farmworkers.

WHAT IS ENVIRONMENTAL JUSTICE?

Environmental justice is a term that encompasses both social and environmental issues. Environmental justice is the achievement of equal protection from environmental and health hazards for all people regardless of race, income, culture, or social class. Therefore, it is implicit that laws must be applied with fairness and impartiality with regard to all socioeconomic, race, income, and geographic differences (Murphy-Greene 2007, 473–90).

Environmental justice refers to the distribution of environmental risks across population groups and to the policy responses to these distributions. The four main areas of environmental justice focus on: (1) the distribution of environmental hazards; (2) the distribution of the effects of environmental problems; (3) the policy-making process; and (4) the administration of environmental protection programs. By focusing on these four areas, one can more easily examine and understand the issue of environmental justice, and as a result, develop strategic methods for addressing the issue.

On the federal level, the United States Environmental Protection Agency (EPA) has taken the lead in the environmental justice movement by developing formal goals. These goals are:

> No segment of the population, regardless of race, color, national origin, or income, as a result of EPA's policies, programs, and activities, suffers disproportionately from adverse human health or environmental effects, and all people live in clean, healthy, and sustainable communities.
>
> Those who live with environmental decisions—community residents, state, tribal, and local governments, environmental groups, businesses—must have every opportunity for public participation in the making of those decisions. An informed and involved community is a necessary and integral part of the process to protect the environment (U.S. EPA 1995a, 1–2).

HOW THE MOVEMENT BEGAN

The Environmental Protection Agency was created in 1970. One year after its creation the Council on Environmental Quality published their annual report acknowledging racial discrimination adversely affects urban poor and the quality of their environment (U.S. EPA 1995b, back of cover page). In 1979 Robert Bullard published a study of an affluent African American community's attempt to block the siting of a sanitary landfill. The environmental justice movement gained recognition in 1982 when the residents of the predominantly black Warren County, North Carolina, protested against the siting of a polychlorinated biphenyl landfill in their county (U.S. EPA 1995b, back of cover page; Bryant and Mohai 1992b, 8; U.S. EPA 1992, 16; Callahan 1994, 88). The Warren County protest set off a chain of protests in the community. These protests triggered an investigation by the General Accounting Office of the socioeconomic and racial composition of communities surrounding the four major hazardous landfills in the southern region (U.S. EPA Region 4) of the United States, which included the states of Mississippi, Alabama, Georgia, South Carolina, North Carolina, Kentucky, and Florida (U.S. General Accounting Office 1983). In 1983, the authors of the GAO study concluded three of the four landfills were located in predominantly black neighborhoods.

In 1987, the United Church of Christ conducted a study in which patterns associated with commercial hazardous waste facilities and uncontrolled toxic waste sites were examined (Goldman 1994, 2–8; Keeva 1994, 90; Rosen 1994, 223; United Church of Christ 1987, xiii). When examining the demographic characteristics of communities with commercial hazardous waste facilities, the study found that: (1) race was the most significant variable tested in association with the location of commercial hazardous waste; (2) communities with the greatest number of hazardous waste facilities had the highest number of minorities; and (3) communities with one commercial hazardous waste facility had twice the average number of minorities compared to those communities without such facilities (United Church of Christ 1987, xiii).

In 1990 several major events occurred which significantly advanced the environmental justice movement. First, Bullard published what is considered by most the first textbook on environmental justice, *Dumping in Dixie* (Bullard 1990, 7–116). Bullard blames the federal government for the "urban apartheid" which exists in the United States. Bullard argues all levels of government are to blame for the institutional racism and discriminatory land-use policies and practices that exist in the United States due to their influence on the creation and perpetuation of racially separate and unequal residential areas for blacks and whites. Two other important arguments of Bullard's book are that environmental discrimination is easier to document empirically than to prove in a court of law, and that Florida was one of the three states in the United States known for not having strong pollution prevention and environmental programs.

Another important event in 1990 was a conference called the Conference on Race and the Incidence of Environmental Hazards hosted by the University of Michigan School of Natural Resources. The conference revealed that data on environmental injustices had been available for over two decades. The conference also focused on the identification of further research areas (Bryant and Mohai 1992b, 8).

The third major event of 1990 was as a result of the Michigan conference. Then–EPA administrator William Reilly created the Environmental Equity Workgroup. The workgroup focused on three tasks: (1) review and evaluate the evidence that racial minority and low-income people bear a disproportionate risk and burden of environmental pollution and hazards; (2) review current EPA programs to identify factors that might give rise to different risk reduction and develop approaches to correct such problems; and (3) review institutional relationships, including outreach to and consultation with racial minority and low-income organizations, to assure that the EPA was fulfilling its mission with respect to these populations (Bryant and Mohai 1992b, 8).

The EPA made further strides in advancing the environmental justice movement in 1992. First, the agency released the report titled *Environmental Equity: Reducing Risk for All Communities* (U.S. EPA 1992, 16). The term environmental equity was the former term used by the EPA to describe environmental justice. This report illustrated how the issue of environmental equity may be approached from several different perspectives, such as region, ethnic/racial group, type of pollutant, and type of illness caused as a result of exposure to pollutants.

Second, the Office of Environmental Equity was established at the EPA (U.S. EPA 1995a; Ember 1994, 22–23). The name was later changed to the Office of Environmental Justice (OEJ). The establishment of the OEJ by the EPA gave the issue of environmental justice increased credibility and legitimacy. Then–agency administrator Carol Browner made environmental justice an agency priority in 1993 (U.S. EPA 1995a). In addition to the establishment of the OEJ in 1993, the EPA also established the National Environmental Justice Advisory Council (NEJAC) through a charter to provide independent advice, consultation, and recommendations to the administrator of the EPA on matters related to environmental justice (NEJAC 1994, 1,2).

Also in 1992 was the publication of a *National Law Journal* article by Coyle and Lavelle titled "Unequal Protection." The article highlighted how white communities see faster action, better results, and stiffer penalties on environmental issues than minority communities (Rosen 1994, 223; Coyle and Lavelle 1992, S1–S2). This article focused on the inequities in the way the EPA enforced its laws. The authors contended there was a clear difference in the way the U.S. government cleaned up toxic waste sites and punished polluters in minority communities and white communities.

Executive Order 12898

On February 11, 1994, President Clinton issued Executive Order 12898, titled "Federal Actions to Address Environmental Justice in Minority Populations and Low-Income Populations" (Clinton 1994, 1). The executive order designated eleven federal agencies to be accountable for environmental justice and stated that as part of the National Performance Review (a federal agency reorganization plan), each federal agency would make achieving environmental justice a part of its mission. Each agency is to accomplish this by identifying and addressing, as appropriate, disproportionately high and adverse human health or environmental effects of its programs, policies, and activities on minority populations and low-income populations in the United States and its territories and possessions, the District of Columbia, the Commonwealth of Puerto Rico, and the Commonwealth of the Mariana Islands.

Also, Executive Order 12898 created the Interagency Working Group on Environmental Justice which is comprised of eighteen executive agencies and offices. The group's members are the Department of Defense, Department of Health and Human Services, Department of Housing and Urban Development, Department of Labor, Department of Agriculture, Department of Transportation, Department of Justice, Department of the Interior, Department of Commerce, Department of Energy, Environmental Protection Agency, Office of Management and Budget, Office of Science and Technology Policy, Office of the Assistant to the President for Domestic Policy, National Economic Council, Council of Economic Advisers, and such other government officials the President may designate. The Working Group's main goals are to: (1) provide guidance to federal agencies on criteria for identifying disproportionately high and adverse human health or environmental effects on minority populations; (2) assist in coordinating data collection; (3) examine existing data and studies on environmental justice; and (4) develop interagency model projects on environmental justice that show evidence of cooperation among federal agencies (Murphy-Greene 2007, 473–90).

Following the signing of the executive order, then–EPA administrator Carol Browner emphasized her support for environmental justice issues in her 1994 testimony before the House Appropriations Committee (U.S. House of Representatives 1994). In her testimony, the former administrator expressed the EPA's support for a new generation of environmental protection which invests in several strategic approaches to environmental protection including, among other things, environmental justice. In identifying how the agency will integrate environmental justice throughout its programs, the administrator stated the agency will examine permitting, grants, data collection and analysis, and enforcement.

Other Significant Events in the Environmental Justice Movement

After Executive Order 12898 was issued, several other events occurred which impacted the environmental justice movement. First, a report was issued in 1994 by the United Church of Christ titled *Toxic Waste and Race Revisited* (United Church of Christ 1987, xiii). This report was a follow-up to the 1987 study by the same organization. The report strengthened the association between race and the location of waste facilities noted in the 1987 study.

Next, the first Interagency Public Meeting on Environmental Justice was held in 1995 at Clark Atlanta University in Atlanta, Georgia. The purpose of the meeting was to provide an opportunity for the public to share concerns and recommend changes in the federal agencies' environmental justice strategies. In 1997 the U.S. Office of Environmental Justice released the *Environmental Justice Implementation Plan*. This plan outlined specifically how the agency would implement the rules identified in Executive Order 12898.

From the late 1990s until the present the EPA has worked to meet the needs of communities in the United States most impacted by environmental justice issues. The agency has held outreach activities and workshops in addition to providing funds directly to impacted communities via grant programs. These include the Small Grants Program and the Community University Partnership Grant Program. Through the Small Grants Program the agency awarded communities across the United States thousands of dollars to address their specific needs.

In 2004 the EPA's OEJ launched a grant program called the Environmental Justice Collaborative Problem-Solving Cooperative Agreement Program (U.S. EPA 2007a). This program is designed to provide direct financial and technical assistance to certain community based organizations. In order to aide community members in addressing environmental justice concerns the EPA developed the Collaborative Problem-Solving Model (CPS Model). The CPS Model has been effective in helping various stakeholders such as communities, industry, academic institutions, and others address local environmental and/or public health issues in a collaborative manner.

Aside from the strides the environmental justice movement has made here in the United States, the movement has broadened to include international environmental justice issues in countries such as Nigeria, South Africa, and Mexico, to name a few (Murphy-Greene 2007).

Environmental Risk Assessment and Minority Communities

Thus far, this chapter has provided a chronological history of the advancement of the environmental justice movement. The following section provides an in-depth examination of the issue of environmental risk assessment.

Measuring environmental risk to all groups in society is difficult (Kraft and Scheberle 1995, 113). Environmental justice focuses on the risk of exposure of minorities and low-income groups. When examining the literature on environmental risk, it is evident that scholars have differing views on the best approach to measure or assess risk. Finkel and Golding (1993, 50) favor the use of a refined comparative risk assessment (CRA) in order to achieve sensible priorities. This refined method actively involves laypeople in the ranking exercises. The two main goals of CRA are risk reduction and risk assessment. However, environmental justice scholar Robert Bullard argues that risk-based priority setting does not always help minority populations. Bullard argues that risk-based priority setting may perpetuate the failure to tackle the true problem. Bullard argues for the Multiple Risk Approach (MRA) which puts priority on all of the geographic areas where minorities and low-income populations face multiple risks from many sources.

Aside from CRA and the MRA, Burns et al. (1992, 137) found that perceptions of risk and social responses were more strongly related to exposure to risk than to the magnitude of exposure. The authors argued that what human beings perceived as threats to their well-being was influenced by their values, attitudes, social influences, and cultural identity. The authors discussed the framework of social amplification of risk, which integrated the technical assessment and the social experience of risk. When examining the possible reasons why minority communities were often overlooked in the risk assessment process, these authors argued that minority groups are socialized to accept a certain level of environmental risk as normal, whereas middle to upper-class, nonminority communities perceive the same level of environmental risk as much more severe. The authors noted that nonminorities and nonpoor were better able to call greater media and government attention to these perceived risks, leading to faster resolution of environmental problems in nonminority and nonpoor communities.

Bullard (1993a, 188) rejects the argument by Burns et al. that minority communities are more accepting of environmental hazards than nonminority communities. Instead, Bullard argues that minorities' lack of social power is the main determinant of where hazardous waste sites are located. Bullard (1993b, 23) also asserts that environmental racism exists within local zoning boards as well as within the U.S. Environmental Protection Agency and nongovernmental organizations such as mainstream national environmental and conservation groups.

Agricultural Workers

The environment has been a politically sensitive topic in the United States since 1962 when Rachel Carson (1962) in her watershed book *Silent Spring* discussed the harmful effects of pesticides. Carson was successful in raising the awareness of federal officials and the general public of the harmful chemicals used to kill insects. Through her efforts the pesticide DDT was banned in the United States. As a result of the banning of DDT the bald eagle was recently removed from the Endangered Species list (Herbert 2007, 3).

The use of pesticides and herbicides by farmers has continued to gain attention in the United States. When examining the harmful effects of pesticides little attention has been given to the impacts those chemicals have on agricultural workers. The EPA defines agricultural workers as individuals involved in the production of agricultural plants (U.S. EPA 2007b). The terms farmworker and agricultural worker will be used interchangeably throughout this chapter. This chapter will now examine the issue of pesticides and farmworkers in Florida.

Farmworkers are one of the most socially, economically, and politically marginalized occupational groups in the United States (U.S. EPA 2007b). Their marginalization is a result of a combination of their inability to speak English, their low-income and education level, and their lack of home ownership. Generally, farmworkers lack the ability to speak and read English and often times they lack the ability to read in their native language. As a result they are unable to read and understand the laws that protect them from environmental hazards. Because of their marginalization, farmworkers are at greater risk of exposure to environmental hazards, such as pesticides.

According to Moses, when examining the distributional impacts of environmental pollution across different racial and income groups in America, agricultural workers face an average risk of skin disease four times higher than workers in other industries (Moses 1989, 120–24; Moses 1993, 167). Moses also notes that more than 40 percent of all reported occupational diseases in the United States were disorders of the skin and the actual incidence was estimated to be ten to fifty times higher than reported. Moses asserts the prime cause of the high rate of skin disease among farmworkers is attributed to pesticide exposure. Agricultural workers suffer from a variety of health problems including spontaneous abortions, still births, low sperm count, sterilization, cancer, neurological and behavioral disorders, and other related illnesses. Many of these health problems go unreported due to the farm workers' lack of access to proper health care. Generally, farmworkers are not provided health care by their employer and are therefore dependent upon community clinics or the emergency room of the local hospital for primary care. Since most farmworkers do not own their own vehicles, transportation to health care providers is another barrier to their receiving proper medical treatment.

THE WORKER PROTECTION STANDARD

When examining the laws and regulations designed to protect farmworkers from the harmful effects of pesticides, the U.S. Environmental Protection Agency's Worker Protection Standard for Agricultural Workers (WPS) is the primary regulation aimed at reducing the risk of pesticide poisoning and injuries among agricultural workers and pesticide handlers (U.S. EPA 2007b). According to the EPA, the WPS offers protections to 3.5 million agricultural workers and pesticide handlers (people who mix, load, or apply pesticides) in the United States today. These workers are spread out over 600,000 agricultural establishments (U.S. EPA 2007c, 5). The WPS specify that growers are required to "restrict entry into treated areas; provide notification of pesticide applications; post specific information regarding pesticide applications (what, where, and when); assure that workers have received safety training;

post safety information; provide decontamination supplies; and provide access to emergency medical assistance when needed" (U.S. EPA 2007c, 5).

Several previous studies have noted weak enforcement of the WPS in Florida (Murphy 2002a, 281–314; Murphy-Greene and Leip 2002, 650–58). Murphy-Greene and Leip conducted a study involving personal interviews with 178 farmworkers in three counties in Florida (Palm Beach, Indian River, and Collier Counties) (Murphy-Greene and Leip 2002, 650–58). Of those farmworkers interviewed for the study, 10 percent reported being sprayed directly with pesticides while they worked in the fields and 64 percent reported that an airplane or tractor sprayed pesticides on the field next to where they were working. This study also reveals 82 percent of the farmworkers interviewed did not know when the fields were last sprayed with pesticides before they reentered the fields. Thus it is evident the WPS are not properly implemented and enforced on these Florida farms. State agencies generally have the primary responsibility for enforcing WPS violations. However, the EPA has jurisdiction in Wyoming and partial jurisdiction in Colorado. Also, the EPA will prosecute cases referred to it by the states.

It is difficult to determine the actual number of farmworkers in the United States due to the migratory nature of the workers. As a result, federal and state counts of farmworkers vary. A state count of Florida migrant and seasonal farmworkers put the number in 1999 at 59,464 workers (Murphy 2002a, 281–314). However, based upon currently available data, Florida is one of the top three agricultural states in the United States along with California and Texas with regard to the number of workers. As such a large agricultural state enforcement of the WPS is a difficult task, as noted in previous studies (Murphy 2002a, 281–314; Murphy-Greene and Leip 2002, 650–58).

It is evident within the last several years the EPA has made increased efforts to enforce the WPS. According to John Peter Suarez, EPA Assistant Administrator for Enforcement and Compliance Assurance, "Environmental justice is one of the highest priorities for EPA's enforcement program, and this Agency will take whatever steps are necessary to ensure agricultural workers and pesticide handlers are protected from harmful exposure to pesticides. . . . The federal government will not tolerate growers who place their workers in harm's way because they fail to comply with the law" (U.S. EPA 2007d, 4).

In 2005 the EPA started making available to the public violations of the Worker Protection Standards that occurred during inspections. The data are broken down into the ten EPA regions. EPA Region 4 includes the states of Alabama, Florida, Georgia, Kentucky, Mississippi, North Carolina, South Carolina, and Tennessee. The categories referred to in the tables are defined in table 7.1.

Table 7.1. Worker Violation Categories*

Pesticide Safety Training	Failure of agricultural employer to ensure that each worker or handler is trained according to the provisions of the WPS.
Central Posting	Failure of agricultural employer to display required information at a central location on the agricultural site where it is readily accessible to workers and handlers.
Notice of Application	Failure of agricultural employer to ensure that notification and posting requirements for pesticide applications comply with the WPS label.
Entry Restrictions	Failure of agricultural employer to ensure that no one enters or remains in the treated area during a pesticide application other than an appropriately trained and equipped handler, until the time specified by label and regulations has elapsed.
Personal Protective Equipment	Failure of agricultural employer to assure that handler uses the clothing and PPE specified on the label for use of the product.
Decontamination	Failure of agricultural employer to provide decontamination supplies for washing off pesticides for workers and handlers in accordance with the WPS.
Emergency Assistance	Failure of agricultural employer to provide emergency medical assistance and prompt transportation to an appropriate emergency medical facility to a worker or handler.
Retaliation	Action taken to punish or discourage a worker or handler from complying or attempting to comply with any requirement of the WPS by the agricultural employer.

Source: U.S. EPA (2007d). FY2005 WPS Inspection and Enforcement Accomplishment Report, Office of Compliance

Note: *This table does not show the two categories "Mix or Loading, Application Equipment" and "Information Exchange" since these categories only apply to pesticide handlers.

Table 7.2 reveals the large number of violations for Region 4. Region 4 has more than three times as many violations of Pesticide Safety Training and twice as many violations of Central Postings than any other region. According to the EPA Office of Compliance, many inspections take place on the state level and are thus not required to be reported to the EPA on the federal level. This makes it difficult for the EPA to obtain a full picture of the true number of violations taking place within each state.

Table 7.3 displays the number of WPS violations during inspection for 2005, 2006, and 2007 for each state within EPA Region 4.

For each year Florida has by far the most violations when compared to any other state in the Region. Most of the violations are in the categories

Table 7.2. Worker Protection Standard Violations during Inspections for 2005, 2006, and 2007: All Regions

Regions	Pesticide Safety Training			Central Posting			Notice of Application			Entry Restrictions			Personal Protective Equipment			Decontamination Supplies			Emergency Assistance			Retaliation		
	'05	'06	'07	'05	'06	'07	'05	'06	'07	'05	'06	'07	'05	'06	'07	'05	'06	'07	'05	'06	'07	'05	'06	'07
Region 1	14	10	16	23	27	14	1	4	0	0	0	0	1	11	6	0	1	2	0	0	0	0	0	0
Region 2	52	88	66	57	57	37	1	68	34	0	5	1	28	9	13	33	74	22	8	43	11	0	1	0
Region 3	13	93	164	12	66	115	7	18	66	1	0	0	4	26	29	4	15	13	0	1	0	0	0	0
Region 4	282	253	320	331	287	324	46	29	51	34	45	23	44	53	79	69	68	106	4	7	2	0	0	0
Region 5	65	101	36	72	112	39	29	32	17	7	9	2	36	28	18	30	43	33	2	2	0	0	0	0
Region 6	18	19	14	13	8	10	2	3	0	4	0	1	2	2	3	6	2	1	2	2	0	0	0	0
Region 7	25	36	31	26	31	21	8	16	17	7	11	8	10	17	13	12	9	8	8	8	7	4	0	1
Region 8	21	32	26	22	41	35	4	9	10	4	4	5	6	11	5	4	18	24	3	3	4	0	0	0
Region 9	37	53	69	29	49	76	22	29	22	6	13	9	20	20	23	16	19	40	2	1	6	0	0	1
Region 10	87	52	62	165	97	96	20	11	20	8	3	9	41	44	40	42	40	45	1	10	2	0	1	0
TOTAL	614	737	804	775	778	767	140	219	237	29	146	58	192	221	229	216	289	294	30	77	28	4	1	2

Sources: U.S. EPA (2007d). FY 2005, 2006 WPS Inspection and Enforcement Accomplishment Report, Office of Compliance; U.S. EPA (2008). FY 2007 WPS Inspection and Enforcement Accomplishment Report, Office of Compliance

Table 7.3. Worker Protection Standard Violations during Inspections for 2005, 2006, and 2007: Region 4

Region 4	Pesticide Safety Training '05	'06	'07	Central Posting '05	'06	'07	Notice of Application '05	'06	'07	Entry Restrictions '05	'06	'07	Personal Protective Equipment '05	'06	'07	Decontamination Supplies '05	'06	'07	Emergency Assistance '05	'06	'07	Retaliation '05	'06	'07
Alabama	24	6	18	49	9	22	0	0	33	0	0	0	3	0	10	1	0	3	1	0	0	0	0	0
Florida	106	110	181	126	109	193	10	13	9	15	25	8	19	24	49	61	45	76	2	0	0	0	0	0
Georgia	40	32	15	76	71	29	0	0	0	16	11	8	4	5	4	2	2	2	1	0	0	0	0	0
Kentucky	17	11	25	10	10	3	5	1	0	2	3	0	3	1	0	1	6	0	0	3	0	0	0	0
Mississippi	3	16	25	2	15	9	0	4	6	1	2	6	0	4	2	0	3	4	0	4	0	0	0	0
N. Carolina	39	57	49	27	68	30	31	1	1	0	3	1	15	18	12	2	10	16	0	0	0	0	0	0
S. Carolina	14	11	20	5	3	25	0	9	0	0	0	0	0	1	2	2	2	4	0	0	0	0	0	0
Tennessee	39	9	10	36	2	13	0	1	20	0	1	0	0	0	0	0	0	1	0	7	0	0	0	0
Total	282	252	320	331	287	324	46	29	69	34	45	23	44	53	79	69	68	106	4	7	0	0	0	0

Sources: U.S. EPA (2007d). FY2005, FY2006 WPS Inspection and Enforcement Accomplishment Report, Office of Compliance U.S. EPA (2008). FY 2007 WPS Inspection and Enforcement Accomplishment Report, Office of Compliance

of Pesticide Safety Training and Central Posting. This means farmworkers are not receiving proper pesticide safety training and there are not signs posted in the fields notifying farmworkers when pesticides were last applied to the fields.

As noted earlier in the chapter, California and Texas are two of the largest agricultural states in the United States. However, as revealed in table 7.4 the number of WPS violations in these two states is considerably lower than in Florida. For 2005, 2006, and 2007 for the two categories Pesticide Safety Training and Central Posting, where the highest number of violations generally occur, Florida has 106 and 126 violations in 2005, and 110 and 109 violations in 2006, and 181 and 193 violations in 2007 respectively. In those same two categories for 2005 California has 5 and 12 violations, in 2006 there are 10 and 22 violations, and in 2007 there are 4 and 10 violations. Texas has 11 and 5 violations in 2005, and 11 and 2 violations in 2006, and 11 and 7 violations in 2007. These low numbers for such large agricultural states may indicate the data is not being reported fully to the EPA and instead is being reported on the state level. States only report WPS violations to the EPA based on the amount of funding provided to each state by the EPA for inspections. Large states like California and Texas are likely to provide much of the funding for inspections at the state level and thus are not required to report the data to the EPA.

In 2005 the EPA began collecting data and publishing a WPS Inspection and Enforcement Accomplishment Report. The data in the report are based on data submitted by states and tribes or the regional program staff in cases where the EPA manages the WPS program. Since some states and tribes take separate action, the data represent only those inspections reported to EPA and may exclude inspections and enforcement actions taken by a state or tribe. Among other things, this report shows the number of warnings issued, administrative hearings conducted, cases forwarded to EPA, and other enforcement actions.

Table 7.5 reveals the large number of WPS inspections and warnings issued in EPA Region 4. From the data collected at the federal level by the EPA it is difficult to make any conclusion about the effectiveness of the Worker Protection Standards since not all state data is reported at the federal level. Therefore some states may have many more violations reported at the state level that are not reported to the federal government. Table 7.6 reveals that Florida has many more inspections and violations than any other state in the region. However, due to the inconsistencies in reporting by the states, it is difficult to draw firm conclusions from these numbers. This study reveals a fragmented system that would benefit from more information sharing between the states and federal government. This would provide the EPA with a better picture of what is happening within each state. While the data do not provide the full picture, one is able to view a snapshot of what is

Table 7.4. Worker Protection Standard Violations during Inspections 2005, 2006, and 2007: California and Texas

	Pesticide Safety Training			Central Posting			Notice of Application			Entry Restrictions			Personal Protective Equipment			Decontamination Supplies			Emergency Assistance			Retaliation		
	'05	'06	'07	'05	'06	'07	'05	'06	'07	'05	'06	'07	'05	'06	'07	'05	'06	'07	'05	'06	'07	'05	'06	'07
California	5	10	4	12	22	10	3	0	0	1	4	0	7	9	3	9	1	10	0	0	0	0	0	0
Texas	11	11	11	5	2	7	0	1	0	0	0	0	0	0	1	3	0	0	0	0	0	0	0	0

Sources: U.S. EPA (2007d). FY 2005, FY 2006 WPS Inspection and Enforcement Accomplishment Report, Office of Compliance; U.S. EPA (2008). FY 2007 WPS Inspection and Enforcement Accomplishment Report, Office of Compliance

Table 7.5. WPS Inspection and Enforcement Accomplishment Report 2005–2007: All Regions

	Region 1	Region 2	Region 3	Region 4	Region 5	Region 6	Region 7	Region 8	Region 9	Region 10	Total
2006*	220	379	137	1688	228	69	94	249	262	197	3523
2007*	271	178	508	2002	161	950	92	235	255	195	4847
2005**	29	45	11	282	52	19	26	33	41	34	572
2006**	46	90	22	284	44	35	44	34	37	63	699
2007**	20	63	23	233	19	16	47	32	39	60	552

Source: U.S. EPA (2008). FY2005, 2006, & 2007 WPS Inspection and Enforcement Accomplishment Report, Office of Compliance

Notes: *Total Number of Inspections, **Total Number of Warnings Issued

Table 7.6. WPS Inspection and Enforcement Accomplishment Report 2005–2007: Region 4

Region 4	Total Number of Inspections*	Number of Warnings Issued	Total Number of Inspections	Number of Warnings Issued	Total Number of Inspections	Total Number of Warnings Issued
	2005	2005	2006	2006	2007	2007
Alabama	—	2	56	56	80	0
Florida	—	59	685	127	996	130
Georgia	—	80	253	47	217	14
Kentucky	—	19	75	15	47	10
Mississippi	—	3	79	0	97	61
N. Carolina	—	80	345	3	451	4
S. Carolina	—	14	78	17	15	0
Tennessee	—	25	117	19	99	14
Total	**—**	**282**	**1688**	**284**	**2002**	**233**

Source: U.S. EPA (2008). FY2005, 2006, & 2007 WPS Inspection and Enforcement Accomplishment Report, Office of Compliance

Note: *Data not available for 2005 for total number of Inspections.

happening in some states with regard to enforcement of the WPS. Florida having 127 and 130 warnings issued for 2006 and 2007 respectfully indicates some problems on the state level with proper implementation of the law. In most regions the number of warning issues for 2007 declined from 2006. This may show some progress. However, in order to obtain a more complete picture of the effectiveness of the WPS, state data would need to be examined.

CONCLUSION

The issue of environmental justice has come a long way since it first began in 1982 in Warren County, North Carolina. The EPA's Worker Protection Standard has made progress by making public the inspection and enforcement data. The simple fact that the data on the WPS violations are now available to the public is progress in providing agricultural workers better protection under the law and thus protection from environmental hazards. This is an incremental step in the right direction for providing the farmworkers with better protection from environmental hazards. One can only hope growers will make better efforts to properly implement the WPS. Enhanced information sharing between the states and federal government would provide a more complete picture of the effectiveness of the WPS.

This chapter has provided an overview of the environmental justice movement from its inception to the present day with an in-depth look at agricultural workers in the South. It appears from the data that environmental justice still does not exist in the South as evident by the high number of violations of the WPS reported to the EPA by the states. While state level data would help provide a better understanding of the effectiveness of the WPS at the state level, the data provided in this study does demonstrate a high number of WPS violations occurring in some states. These violations point to the lack of protection from environmental hazards agricultural workers are provided. With continued efforts in enforcement and enhanced information sharing by the state governments, the states in the South have the ability to correct the current situation and provide farmworkers a safer work environment.

One of the reasons the South, and particularly Florida, may have such a dismal record when it comes to environmental justice is what Elazar (1984) refers to as political culture. Southern states are characterized by Elazar as traditionalistic, meaning they hold on to traditions and values. Within the South, states' political culture may vary as do Florida's and North Carolina's. While Florida is characterized as traditionalistic-individualistic, North Carolina is characterized as traditionalistic-moralistic. The individualist states typically place a higher value on the rights of individuals, while moralistic states place a higher value on the greater good of society. This might help explain why North Carolina played such a large role in shaping the environmental justice movement starting with the Warren County, North Carolina, protest in 1982. North Carolina being a traditionalistic-moralistic state places a high value on tradition, yet also demonstrates a high value for the greater good of society, thus possibly shedding some light on why North Carolina is considered by many the birthplace of environmental justice. Florida, on the other hand, as a traditionalistic-individualistic state places a high value on tradition and the rights of individuals, possibly explaining why Florida has such a poor record when it comes to protecting agricultural workers. Florida appears to place a higher value on the rights of the growers than the rights of the agricultural workers.

This chapter highlights the need for a comprehensive study of individual state enforcement data of the WPS. Data collected at the state level needs to be compared with the data provided to the EPA by the states. A comparative analysis of this data would shed more light on the true effectiveness of the WPS.

REFERENCES

Bryant, B., and P. Mohai.1992a, Race, Poverty, and the Environment. *EPA Journal*, 18:8.
———. 1992b. The Michigan Conference: A Turning Point, *EPA Journal* 18:10.
Bullard, R. 1990. *Dumping in Dixie: Race, Class, and Environmental Quality.* Boulder, CO: Westview Press.
———. 1993a. *Confronting Environmental Racism: Voices from the Grassroots.* Boston: South End Press.
———. 1993b. The Threat of Environmental Racism. *Natural Resources & Environment.* Winter:23.
Burns, W., et al. 1992. The Social Amplification of Risk: Theoretical Foundations and Empirical Applications. *Journal of Social Issues* 48:137.
Callahan, P. 1994. Environmental Racism: When Civil Rights Are Used to Protect More Than Individual Liberty. *Omni* 16:88.
Carson, R. 1962. *Silent Spring.* Cambridge, MA: The Riverside Press.
Clinton, W. 1994. Executive Order 12898 of February 11, 1994: Federal Actions to Address Environmental Justice in Minority Populations and Low-Income Populations. *Federal_Register* 59:1.
Coyle, M., and M. Lavelle. 1992. Unequal Protection. *National Law Journal* 15:S1.
Elazar, D. 1984. *American Federalism: A View From the States.* 3rd ed. NY: Harper and Row.
Ember, L. 1994. EPA Giving Increased Priority to Environmental Issues. *Chemical & Engineering News* 72:22–23.
Finkel, A., and D. Golding. 1993. Alternative Paradigms: Comparative Risk Is Not the Only Model. *EPA Journal* Jan./Feb./March:50–52.

Goldman, B. 1994. Not Just Prosperity: Achieving Sustainability with Environmental Justice. *National Wildlife Federation: Corporate Conservation Council.* 2–8.

Herbert, H. J. 2007. A Mightly Comeback. *The Virginian Pilot.* June 28:Sec. A.

Keeva, S. 1994. A Breath of Justice. *ABA Journal* 80:90.

Kraft, M., and D. Scheberle. 1995. Environmental Justice and the Allocation of Risk: The Case of Lead and Public Health. *Policy Studies Journal* 23:113.

Moses, M. 1989. Pesticide-Related Health Problems and Farm Workers. *AAOHN Journal.* 37:120–24.

———. 1993. Farm Workers and Pesticides. *Confronting Environmental Racism: Voices From the Grassroots.* ed. Robert Bullard. Boston: Southend Press.

Murphy, C. 2002a. The Occupational Safety and Health of Florida Farm Workers: Environmental Injustice in the Fields? *Journal of Health and Human Services Administration* 25:281–314.

———. 2002b. Environmental Justice: A Case Study of Farm Workers in South Florida. *International Journal of Public Administration* 25:193–220.

Murphy-Greene, C. 2007. Environmental Justice: A Global Perspective: In *Handbook of Globalization and the Environment.* ed. Khi Thai, Dianne Rahm, Jerrell Coggburn. 473–90. Boca Raton: CRC Press.

Murphy-Greene, C., and L. Leip. 2002. Assessing the Effectiveness of Executive Order 12898: Environmental Justice for All? *Public Administration Review* 62:650–58.

National Environmental Justice Advisory Council (NEJAC). 1994. Proceedings of the National Environmental Justice Advisory Council and Subcommittees: A federal advisory committee. Washington, DC. 1–2.

Rosen, R. 1994. Who Gets Polluted. *Dissent* 41:223.

United Church of Christ. 1987. *Toxic Waste and Race in the United States.* The Commission for Racial Justice, xiii.

U.S. Environmental Protection Agency. 1992. Environmental Equity, Reducing Risk for All Communities. *Volume 1: Workgroup Report to the Administrator: EPA report.* May:18

———. 1995a. Environmental Justice Strategy: Executive Order 12898.

———. 1995b. Environmental Justice 1994 Annual Report.

———. 2007a. EJ Collaborative Problem-Solving Cooperative Agreements Program www.epa.gov/compliance/resources/publications/ej/grants/cps-manual-12-27-06 .pdf (accessed June 26, 2007).

———. 2007b. Worker Protection Standard for Agricultural Pesticides. www.epa .gov/agricultur/twor.html.

———. 2007c. Worker Protection Standards (WPS): Agriculture-Related Enforcement Cases www.epa.gov/agriculture/wpsenf.html#floridawps.

———. 2007d. FY2005 WPS Inspection and Enforcement Accomplishment Report, Office of Compliance.

———. 2008. FY 2007 WPS Inspection and Enforcement Accomplishment Report, Office of Compliance.

U.S. General Accounting Office. 1983. 1980 Data for Census Area where EPA Region IV Hazardous Waste Landfills are Located. Table 4.

U.S. House of Representatives. 1994. Statement of Carol Browner, Administrator, U.S. Environmental Protection Agency before the House Appropriations Committee.

8

A State Government Faces Environmental Management Change

Mississippi's Department of Environmental Quality

Gerald Andrews Emison

INTRODUCTION

Southern state governments share with those of other regions a landscape in which environmental protection is rapidly changing. The command and control system is being supplemented with an incentive-based government management approach (Durant et al. 2004). At the same time the federal government is reducing its role in environmental protection in favor of more state-based regulatory decisions. The consequences of these developments place state environmental protection methods under change that is unparalleled since the modern environmental era began in the 1970s.

While the South faces conditions similar to those elsewhere, navigating such change may be especially difficult for southern states. The traditionalistic culture that predominates in the South (Elazar 1984) discourages departure from conventional approaches by government. Fluid adaptation to new conditions is unlikely given southern states' cultural orientation, yet such adaptation will be necessary as environmental protection methods and institutional responsibilities change. Southern state governments were tardy in moving into the modern environmental era (Ringquist 1993), and now after making this move, they face further changes necessitated by emerging environmental protection approaches (Fiorino 2006).

In addition to cultural barriers, the state governments' resource positions in the South pose further challenges. Southern states are among the most frugal in allocating funds to public activities (Ringquist 1993). This results in states that have limited financial flexibility to take on the new requirements associated with changing approaches to environmental protection.

These circumstances create a unique challenge for southern states' environmental agencies.

Among southern states, Mississippi represents the most traditionalistic political culture (Elazar 1984). Also it falls near the bottom of rankings of environmental expenditures (Ringquist 1993). As a result, Mississippi presents a sharply drawn example of a southern state government facing the challenges of emerging environmental protection practices. Understanding the conflicts and opportunities Mississippi faces in this turbulent situation can yield insight into the challenges faced by other states, especially southern states, as they migrate from the command and control regulatory system to a more mixed-methods approach to environmental protection.

This chapter examines the barriers and opportunities faced by the Mississippi Department of Environmental Quality (MDEQ) under these emerging conditions through the views of those who are involved in steering public environmental organizations through this period. Through a transect from state to region to the national level, it examines the views of government environmental executives. In addition it examines the views of interest groups representing those state environmental agencies at the national level, identifies conditions under which MDEQ may be more effective in the future, and identifies more general circumstances applicable to other state environmental agencies. In doing this we seek to identify uniquely southern challenges and responses to changes in environmental protection methods that may be relevant to other states.

EMERGING SHIFTS IN ENVIRONMENTAL PROTECTION IN THE UNITED STATES

The nation has achieved substantial environmental improvement by employing the command and control system in which government establishes regulations for pollution sources to meet, monitors adherence to these regulations, and takes legal action against those sources not complying with regulations. This approach worked well when there were a small number of well-defined sources, the control technologies for the emissions of the sources were well understood, and compliance could be easily monitored. As a result of past success with such sources the characteristics of emission sources of the future that must be controlled today are quite different from those previously. The sources present a different nature of environmental control problem. Modern environmental progress depends upon emission reductions from a large number of small sources that are geographically dispersed and do not respond well to the command and control system. These changed conditions call into question the continued use of the command and control system alone. For example, previously sewer outfalls were an emphasis of regulation; today future water pollution controls focus on agricultural and urban runoff.

At the core of environmental management lies state government implementation of federal government standards. The performance of state environmental agencies has been the linchpin that has translated national policy into tangible environmental gains under the command and control system. State governments have evolved institutional structures as well as management processes to successfully deploy the command and control system. In short, state environmental agencies have become well suited to apply techniques that are unlikely to be effective in the future. The nation's continued success in environmental protection depends upon state governments identifying and adopting techniques and practices that are not currently part of their repertoire.

The change in the nature of pollution sources and control strategies is occurring simultaneously with a change in the landscape of intergovernmental relations: The federal government is devolving responsibility from the national level to the state level. Such devolution is proceeding at an accelerated pace as the U.S. Environmental Protection Agency (EPA) seeks to delegate programs it originally operated. Whether termed devolution, unfunded mandates, or abdication of responsibility, this change in intergovernmental relations places increasing stress upon state governments' management capacities at a time in which they must make fundamental changes in how they manage in order to protect the environment.

Public management scholars, environmental practitioners, and the regulated community agree on the growing importance of market-based incentives, information technologies, flexibility, local control of pollution control activities, and the use of techniques other than legal enforcement actions. This need for augmenting the command and control system has been widely acknowledged for over ten years. At Congress' direction the National Academy of Public Administration (NAPA) undertook in the 1990s a series of studies that examined the circumstances propelling the need for change. Initially, a panel of NAPA fellows identified the conditions at the national level that prompted the need for moving from the single-method command and control system to a more mixed methods approach for environmental management (NAPA 1995). This work concentrated on the internal management of EPA and changing EPA's organizational culture. In a follow-up book another NAPA panel examined the emerging stresses on relationships among EPA, the states, and local communities (NAPA 1997). This research provided a series of case studies that documented the need for community-based environmental protection, market-based instruments, public access to environmental information, and other forms of effective voluntary compliance. The report of NAPA's most recent panel on the subject described how EPA might alter its traditional programs to incorporate some of the recommendations of previous panels (NAPA 2000). Concurrent with NAPA's work there have been other parallel streams of analysis and recommendations examining the question of changing the national environmental management system. The National Environmental Policy Institute (NEPI 1997), which is generally associated with a market-based orientation, the Enterprise for the Environment (Enterprise for the Environment 1998), associated with the Georgetown University Center for Strategic and International Studies, as well as a number of policy and legal scholars (Davies and Mazurek 1998; Rosenbaum 2008; Schroeder 2000) examined the situation and reached many of the same conclusions that the NAPA panels had reached.

Previous literature has focused on federal government activities in environmental protection as it changes; however, the changes facing the state governments have not been closely examined. This parallels earlier studies of environmental protection programs. Initially the field dwelled upon the federal roles, their effects, and their effectiveness (Wood and Waterman 1994; Rosenbaum 2008; Plater et al. 1998; Scheberle 1997). This attention broadened as state environmental agencies became more prominent players in environmental protection activities. Lowry (1992) profiled state governments' environmental program expenditures by employing the states as the unit of observation. Ringquist examined state governments' environmental program expenditures and their relationships to environmental effectiveness by considering the states in a longitudinal study (Ringquist 1993). This paralleled Lowry's examination of state governments' effectiveness by examining variation across the states to provide insight into state governments' behaviors concerning environmental protection. Bacot and Dawes (1997) also examined state environmental commitment. These analyses of state environmental agencies centered on state environmental agency behavior under the stable command and control approach. These studies, while important in advancing our understanding of state governments as providers of environmental protection, did not closely examine the role of regional variation in environmental management approaches, and especially did not do this while taking into account the changing nature of environmental protection.

Durant et al. (2004) and Fiorino (2006) have drawn attention to the implications of moving from a sole command and control approach to one employing mixed methods such as incentives and the use of information sharing. John (1994) stressed the ascendancy of nonfederally driven environmental protection as an emerging trend. These works sharpened the understanding of the type of activities involved in non–command and control approaches; however they, too, did not consider regional differences. These differences are likely to become quite important as states tailor new approaches to suit the political cultures, economies, and environmental protection values associated with a finer grained environmental protection approach.

This issue is exceptionally important because state environmental agencies are the critical link between federal ambition and local capabilities. Existing federal environmental legislation expects states to assume delegation of established programs, yet we lack insight into how the states might take advantage of emerging opportunities and requirements for incentive-based, locally directed environmental quality management. In particular, environmental management lacks characterization for state governments of the extant circumstances related to the new responsibilities and opportunities emerging. Without such understandings we risk neglecting a key institutional resource, the state governments. States do not have firm pictures of new conditions, and they have a major commitment to the existing command and control regulatory program structure. This is particularly important because we know from established motivational literature in public organizational behavior that those who are not included will likely impede rather than assist future changes.

Such circumstances are especially problematic for southern states considering the traditionalistic culture that predominates in the region. Key (1949), Elazar (1984), and Erickson et al. (1993) identified that southern states hold characteristics that distinguish this region from other areas of the nation. Key (1949) identified the states of the South as distinctive from the rest of the nation yet each state in the region both shared traits with its neighbors while simultaneously being different from neighboring states. Key's contribution to this research was identifying southern distinctiveness as important to the nature of the region.

Elazar (1984) examined the regions of the nation and identified common cultural and political attributes within these regions. He identified the South as presenting a traditionalistic culture that values stability, authority, and continuity as common cultural values that influence the politics of such states. Elazar's findings are particularly relevant since he noted properties of the South that may be important to governmental adaptation.

Erickson et al. (1993) examined state policy and political alignment across the nation and found that policy preferences frequently reflect political alignment of the electorate. Variation among the states, according to Erickson et al., frequently reflects regional patterns of culture reflected through political alignment and manifested in policy choices.

These works on regional distinctiveness confirm that the South is different from other states. They also allow us to identify properties of southern states that could be directly relevant to state government adaptation to changes in environmental management practices. In short the South likely will have a difficult time adapting to rapid and significant shifts in environmental protection of the type faced in changing from a command and control approach to a more mixed methods environmental management.

Government executives are those most likely to face this shift first. They are charged with carrying out environmental protection legislation by designing the plans, issuing the permits, and following up on enforcement of air, water, and hazardous waste pollution regulations. Understanding government executives' actions is key to understanding the responses to such changes in the distinctive southern setting. As a consequence, this research examines a transect of government executives from the state to the region to the national level in order to understand adaptation to change in environmental management facing southern states.

DATA AND METHODS

This research seeks to uncover the conditions facing the Mississippi Department of Environmental Quality (MDEQ) as it adapts to the changing politics, engineering, management, and organizational circumstances that are transforming environmental protection at the state level, especially under the unique conditions of the South. The motivating objective of this research is to advance understanding of a state department of environmental protection in the South for its potential to take advantage of change in environmental management. We seek to develop insight into the issues concerning state environmental agency adaptation that those directly managing these agencies perceive as important. These are the most critical institutions in achieving governmental performance in environmental protection. Exploratory research methods are chosen in order to provide broad coverage while allowing for discovering unexpected insights.

The interview subjects were selected to obtain a cross section of governmental actors' views on changes occurring in environmental policy at the MDEQ level. As discussed above, this group constituted the primary locus of responsibility for responding to changes in state environmental protection approaches and was judged to be essential for identifying the emerging conditions relevant to the research. Senior general management executives at the administrator and deputy administrator levels were interviewed at MDEQ in Jackson, Mississippi, and at the U.S. Environmental Protection Agency's (EPA) Atlanta, Georgia, regional office. In addition senior program managers in EPA's headquarters in Washington, DC, were interviewed. To supplement state views, interviews also were conducted with the executive director of the Environmental Commissioners of the States, the executive director of the Association of State and Territorial Air Pollution Control Administrators/Association of Local Air Pollution Administrators (now the National Association of Clean Air Agencies), and the executive director of the Association of State and Interstate Water Pollution Control Administrators.

The data were collected through fourteen semi-structured interviews conducted June to October 2004. In MDEQ and EPA's Atlanta regional office the interviewees were the senior regulatory officials of each organization.

In EPA's headquarters office a cross section of six media program and staff program executives were interviewed. The governmental manager interest groups were represented by the executive directors of each group concerned with general state environmental protection, air pollution, and water pollution media programs. All of the federal government interviewees were either the senior career official or the senior politically appointed official in the particular organization. For the state agencies interest groups, the executive director was interviewed.

Semi-structured interviews were employed to permit data to be obtained while encouraging the interviewees to offer nuanced insight unavailable through structured questions. Each subject was interviewed for approximately two hours. The interview instrument queried each subject on a sequence of topics.

Each interviewee was asked to identify the strengths that MDEQ brought to each of its relationships with clients of business, local governments, the general public, and EPA. Next the interviewees were asked to identify the strengths that each of these clients brought to its relationship with MDEQ. Problematic features of each of the MDEQ clients' relationships were sought from the interviewees. Then the interview subject was asked to identify the single greatest opportunity and single greatest threat they felt MDEQ faced. Lastly each was provided the opportunity to offer any critical features not covered elsewhere and suggest the single thing they would change between MDEQ and each of its clients. The results were aggregated, interpreted, and analyzed for common patterns and unique insights as described in the following section.

FINDINGS

The findings from this research are grouped into the potential contributions the actors may bring to environmental improvement, the possible problems for such improvement, and the opportunities each actor has for enhancing the environmental management system during the period of change.

Contributions of the Mississippi Department of Environmental Quality

All interviewees identified, directly or indirectly, a range of strengths that MDEQ, or state environmental agencies in general, brought to the changing environmental management situation. Knowledge of the specific situation surrounding an environmental decision was identified as a principal strength. Characteristics of political, economic, institutional, and physical conditions surrounding location-specific environmental issues were viewed as best understood by the state environmental agency. Responsiveness and capability to innovate were lesser but companion strengths. This specific knowledge was important to one respondent because "states are the primary contact on environmental issues for the public." MDEQ's candor with business and the public was an important feature of the agency's performance, and such candor led, in part, to substantial expression of respect by EPA: "MDEQ is staffed with motivated and knowledgeable, creative folks who have very limited resources," according to a senior EPA executive. Being a guarantor to protect the public's interest in environmental protection further reflected this positive relationship with the public and other stakeholders. In particular, the state agency's ability to serve as this guarantor concerning local government environmental actions, as a surrogate for EPA, was identified as a strength.

Contributions from Other Actors

A range of public and private actors have a stake in state environmental management agency success, and these actors' potential contributions were highlighted by many interviewees, regardless of governmental level. For private actors, business interests in Mississippi had substantial political and social influence they could wield to advance or retard environmental quality. In particular the technical expertise of large industries and the ability to innovate of small businesses appeared to be especially important. Private individuals and the general public were frequently identified as important to environmental program success: "They're the key to long range success in terms of support and funding through politics." Of particular note was the unique strength of such individuals to speak truth to governmental power: "When everyone thinks you're wrong, maybe you are." This positive view of public involvement was tempered by concern, however. A number of respondents expressed the view that the public had lost sustained interest in environmental protection: "The public took the environment to the emergency room and then went home."

Both local governments and EPA were identified as possessing considerable potential to act in a positive manner to improve environmental quality. With a deep knowledge of political and social conditions at the local level, combined with the capacity to tailor specific, unique responses to environmental problems, local governments were viewed as presenting unique capabilities. EPA also had unique capabilities, however they were of a different nature. The ability to serve as a convener of stature, when combined with extensive technical knowledge, gave EPA potential to contribute to the state agency's success. Federal grant funds for state program operation along with the "gorilla in the closet" threat of federal action if state action was not satisfactory were also EPA strengths. These contributions were tempered by views that such influence today was stronger on the information side rather than the policy side and on the vision side rather than detailed execution. Respondents frequently saw EPA as excessively constrained by politics in specific operational circumstances.

Problems and Threats to Progress

MDEQ's greatest threat to its ability to adapt and succeed was consistently identified as inadequate resources. Almost all respondents identified state budget circumstances as severely limiting MDEQ's actions. In particular, the impact of budget limitations on agency capacity was significant: "Their (MDEQ's) ability to maintain a core professional state environmental agency is the major problem given Mississippi's budget situation" according to one EPA official. Also, a number of respondents were concerned that the state legislature did not have a well developed appreciation of the importance of environmental protection both to public health and to long-term industrial advancement. This limitation was traced, in part, to values and also to the extraordinary and increasing complexity of many environmental programs.

Most respondents saw formidable barriers to change derived from other actors. A routine problem presented by business interests was that they emphasized their short-term self-interest. This was especially critical when it was combined with business interests perceived limited ability to engage complex regulations that lie at the heart of modern environmental protection: "They have a total lack of understanding of rules." "Industry (in Mississippi) has yet to acknowledge the legitimacy of environmental controls." Such views by industry pose barriers to change that may be quite difficult for MDEQ to surmount.

Barriers also were identified on the part of the general public. Almost every respondent expressed concern about the public's understanding of the scientific principles in play in environmental protection. In particular, public understanding of the nature of environmental risk and how such risk can be managed was repeatedly raised. When this ignorance was combined with limited understanding of governmental restrictions and skepticism of the purposes of governmental actions, a picture of diminishing public influence on environmental management emerged. Of particular importance was the similarity of observations expressed concerning environmental groups at the state and local level. Most interviewees in one way or another expressed the concern that such groups were more interested in advancing their own particular agenda rather than environmental problem solving. This can present particular barriers to change when such narrow interests are advanced. However, this interpretation must be tempered by acknowledging potential response bias from the governmental perspectives.

Local governments' short-term views on planning and land use were of special concern given the likely importance of such activities to future environmental issues. "Locals can't get past parochial need for self-determination." For EPA, respondents expressed a wide range of concerns. This is not surprising given EPA's central role in environmental policy over the past thirty years. In one way or another most respondents expressed concern over a decline in EPA's leadership on environmental issues. "There is a lack of political leadership nationally and in the states." This leads to "a tumultuous political climate." Respondents identified that EPA was losing skilled staff and that its mandate was declining as the important issues shifted from regulation to incentives. "The federal government is walking away from its responsibility to clean up the environment. It has withdrawn the bully pulpit, money, technical support and so on." This, along with EPA's reliance on narrow, "stove-piped" program organization has undermined the agency's ability to respond and often results in EPA "mumbling" when states ask important policy questions. In general, respondents identified EPA's decline as a significant threat to MDEQ's potential success, and they traced such a decline to "incipient irrelevancy" on EPA's part.

Opportunities for Improvement

Despite these difficulties, every person interviewed believed that considerable opportunities existed for advancing environmental quality in Mississippi. The emerging importance of water quantity to influence water quality issues was identified as a considerable opportunity by a number of respondents. "Water is a potential strategic advantage." Also the opportunity to educate the public on the role of risk in environmental decisions could allow development of a better informed, more sophisticated public. This was viewed as creating considerable opportunities to educate the public on the difficult choices likely to be encountered in environmental protection in the next ten years.

A number of respondents went out of the way to express views on EPA's opportunities to assist the state. They simultaneously sought for EPA to reduce its control over state activities ("Send MDEQ the money and get out of the way") while upholding national standards ("EPA must describe what the acceptable standards are"). With states having a strategic advantage in their abilities to innovate, many saw EPA's role changing to stress national leadership, yet non-EPA interviewees consistently expressed doubt that EPA was up to the challenge: "We now know how to solve most environmental problems, but lack the money and leadership to address these." This was often because "states are buffeted by swinging political emphasis, and EPA's almost too stable because its statutes remove flexibility." To one respondent this was because "the precision of our ability to conceptualize solutions has exceeded our ability to produce the solutions."

What to Change

Respondents were asked to consider both problems and opportunities and then identify the single thing they would change about each of the various actors. Few respondents could restrict themselves to a single improvement, however some consistent themes emerged, and these themes matched previous responses.

For MDEQ, most respondents consistently focused on the need for adequate resources. It appears that its impressive talent, savvy, and organizational skill are constrained, in the view of the respondents, by inadequate resources. "MDEQ is doing a good job considering the resources they have." Identification of inadequate resources was not surprising. The vehemence and breadth across all interviewees was compelling, however.

Business should take a longer range view of decisions, and decisions by the CEOs of such businesses themselves should make the decisions based on a suite of business criteria rather than relying on legal advice as the primary driver of environmental decisions. When combined with a richer understanding of scientific and technical issues, many felt improved business decisions would result.

Improving the public's scientific literacy was routinely expressed as an important need. "The public's intellectual understanding of environmental issues and how the issues affect them is very limited." As a result "the general public's lack of science knowledge is increasingly a national problem."

For EPA, respondents asked for a range of changes, often reflecting the particular views of the interviewee. Requests ranged from establishing a physical presence in Mississippi to reorganizing EPA to place enforcement in the pollution media programs. In a number of instances people wished for less partisanship from EPA, expressing concern at the national level about the effects of pork barrel politics and the exercise of partisan views in environmental policy. Another recurring theme was the wish that EPA would change its mindset to one of an equal partner with the state environmental agencies, rather than one of principal versus agent.

CONCLUSION

The respondents identified a number of opportunities for MDEQ as a result of the changing landscape in environmental management. A number of the respondents went out of their way to stress that these opportunities are not confined to Mississippi, but could apply to most other state environmental agencies, especially those of other southern states. A consistent strength identified was locality and situation specific knowledge the state agency has. As environmental management moves into a world of more tailored regulatory circumstances, national or regional policy orientation is expected to diminish in importance and be replaced with site-specific knowledge. MDEQ has a strong role to play in such circumstances. Similarly, the opportunity and capability to innovate was broadly identified as a MDEQ strength. A number of interviewees focused on the increasing role water quantity will play in water quality issues and stressed that circumstance-specific knowledge and action will be of increasing importance. In a related vein, this situation-specific orientation also contributed to another positive feature of the evolving environmental management regime, the capacity of businesses to innovate based on unique knowledge of their industry and the positive atmosphere in Mississippi toward business. When this is paired with EPA's technical knowledge of specific environmental issues, a broad industry-government coalition for innovation may be possible.

These circumstances are proscribed by some very real and difficult conditions. Almost every respondent identified constraints on state resources as likely to impede taking advantage of the identified opportunities. Most interviewees were doubtful that MDEQ, and in fact most other state environmental agencies, would have adequate resources to do much more than provide basic functions, much less achieve innovations, in the future due to state budgets. This was accompanied by considerable concern about other stakeholders' willingness and abilities to respond to potential opportunities. While perhaps self-serving, this nevertheless yields insight into governmental managers' views on barriers to success. Business continues to be viewed as employing a narrow understanding of self-interest. This stifles both innovation and willingness to take risks, whether the risks derive from collaboration with government or adopting new technical approaches. Interestingly, such a cautious attitude also was identified among environmental groups. Many respondents believed that environmental groups engaged environmental issues with a primary motivation to preserve a command and control regulatory scheme while opposing significant collaborative, innovative departures from this approach. When these characteristics of the two primary interest groups in environmental protection are combined with a broadly identified lack of scientific literacy on the part of the public, most respondents did not believe there were large opportunities to take advantage of the emerging changes in environmental management. To overcome such conditions, interviewees felt a strong federal government push would be necessary, and no one anticipated that EPA, the primary federal pollution control actor, would have the political interest to exert a dominating leadership role.

These circumstances present a picture of opportunities that are severely constrained. A clear limitation on this observation concerns the interviewees. All were either government regulatory officials at the state or federal level, or interest group executives who represent such state officials. Based on the interviews conducted of a group who bear primary responsibility to implement environmental programs, Mississippi, and perhaps a larger group of state environmental agencies, faces a world in which conditions are changing, but its ability to adapt is significantly constrained.

REFERENCES

Bacot, H. A., and R. A. Dawes. 1997. State Expenditures and Policy Outcomes in Environmental Program Management. *Policy Studies Journal* 25:355–70.

Davies, J. C., and J. V. Mazurek. 1998. *Pollution Control in the United States: Evaluating the System*. Washington, DC: Resources for the Future.

Durant, R. F., D. J. Fiorino, and R. O'Leary. 2004. Introduction. *Environmental Governance Reconsidered: Challenges, Choices and Opportunities*, Cambridge: MIT Press.

Elazar, D. J. 1984. *American Federalism: A View from the States*. New York: Harper & Row.

Enterprise for the Environment. 1998. *E4E: The Environmental Protection System in Transition: Toward a More Desirable Future*. Washington: Center for Strategic and International Studies.

Erickson, R. S., G. C. Wright Jr., and J. P. McIver. 1993. *Statehouse Democracy: Public Opinion and Policy in the American States*. New York: Cambridge University Press.

Fiorino, D. J. 2006. *The New Environmental Regulation*. Washington: CQ Press.

Key, V. O. 1949. *Southern Politics*. New York: Vintage Books.

John, D. 1994. *Civic Environmentalism: Alternatives to Regulation in States and Communities*. Washington, DC: CQ Press.

Lowry, W. R. 1992. *The Dimensions of Federalism: State Governments an Pollution Control Policies*. Durham, NC: Duke University Press.

National Academy of Public Administration. 1995. *Setting Priorities, Getting Results: A New Direction for EPA*. Washington: NAPA.

———. 1997. *Resolving the Paradox of Environmental Protection: An Agenda for Congress, EPA and the States*. Washington: NAPA.

———. 2000. *Environment.gov: Transforming Environmental Protection for the 21st Century*. Washington: NAPA.

National Environmental Policy Institute. 1997. *Environmental Goals and Priorities: Four Building Blocks for Change*. Washington: NEPI.

Plater, Z. J. B., R. H. Abrams, W. Goldfarb, and R. L. Graham. 1998. *Environmental Law and Policy: Nature, Law and Society*. St. Paul, MN: West.

Ringquist, E. J. 1993. *Environmental Protection at the State Level*, London: Sage.

Rosenbaum, W. A. 2008. *Environmental Politics and Policy*. Washington, DC: CQ Press.

Scheberle, D. 1997. *Federalism and Environmental Policy*. Washington, DC: Georgetown University Press.

Schroeder, C. H. 2000. Third Way Environmentalism. *Kansas Law Review* 48:1–48, Lawrence: University of Kansas Law School.

Wood, B. D., and R. W. Waterman. 1994. *Bureaucratic Dynamics: The Role of Bureaucracy in a Democracy*. Boulder, CO: Westview.

9

Collaborative Environmental Management within a Traditionalistic Political Culture

An Unconventional Approach to Resolving "Wicked" Problems

Madeleine W. McNamara

As wicked problems,[1] fiscal stresses, and resource shortages challenge the capabilities of public organizations, it seems inevitable that the protection and management of environmental landscapes will rely on multiorganizational partnerships. At a time when public organizations are asked continuously to do more with less, collective action may better prepare administrators to respond to environmental problems, implement policy, and strengthen resource protection. Partnerships form as government agencies and nongovernmental organizations interact across bureaucratic boundaries to couple multiple funding streams, resources, and expertise (Imperial 2005; McNamara 2008; Weber 2009). While other approaches are discussed in the literature (see, for example, Fiorino 2006; Rosenbaum 2005), this chapter focuses on a collaborative approach to environmental management and introduces specific techniques that managers can use to address complex environmental problems.

Collaboration is often used to describe long-term and highly integrated interactions within multiorganizational arrangements (see, for example, Agranoff 2006; Bryson, Crosby, and Stone 2006; Thomson and Perry 2006). Collaborative interactions occur when multiple organizations share responsibility for interconnected tasks and work together to pursue collectively complex goals that cannot be accomplished by a single organization (Keast, Brown, and Mandell 2007; Mattessich, Murray-Close, and Monsey 2001; Thomson and Perry 2006; Weber 2009). While this type of interaction can be very beneficial, collaboration typically requires great levels of commitment and time as stakeholders within a particular arrangement interact frequently to develop shared norms, rules, and processes used to make collective decisions impacting mutual interests (Keast, Brown, and

Mandell 2007; Thomson and Perry 2006; Wood and Gray 1991). Therefore, it may be most appropriate to embrace collaboration in situations where other types of interactions were unsuccessful, situations of crisis exist, organizations share responsibility for the problem, or organizations have similar missions (Daley 2008; Huxham 2003; McNamara 2008).

In these arrangements, public managers do not have authority to command a specific course of action. Instead, the collective group establishes goals that are intended to resolve problems that cannot be resolved by an individual organization (Agranoff 2006; McGuire 2006). Rather than formal authority based on position, all participants may lead and mobilize resources in order to attain the objectives of the collective group (Crosby 1996). Flexibility, shared power, and diverse perspectives are embraced. Collaborative arrangements are typically convened by a referent organization (Bryson, Crosby, and Stone 2006; Wood and Gray 1991). A convener identifies an important problem and plays a significant role in bringing legitimate stakeholders together to achieve a particular purpose (Bryson, Crosby, and Stone 2006; Wood and Gray 1991). Through resource and information exchange, a referent organization facilitates interactions between organizations and generates stability within the organizational environment (Morris and Burns 1997). Since personnel within the convening organization do not have formal power or authority over other organizations within the arrangement, informal influence must be generated through expertise and credibility (Keast et al. 2004; Wood and Gray 1991).

While a command and control approach to implementing environmental policy may be expected in states with a traditionalistic political culture (Elazar 1984), complex problems in Virginia's coastal areas were addressed through collaboration. This case study provides one example in which organizations worked within a traditionalistic political culture while moving beyond a command and control approach to protect and enhance the environment. What collaborative techniques can environmental managers use to address complex problems? How can collaborative interactions be utilized in southern states with a traditionalistic political culture? These questions are addressed in this chapter by exploring an environmental program implemented in Virginia. The first section of this chapter describes the study's setting—the Virginia Coastal Zone Management (VCZM) Program. This section is followed by an explanation of the methodology used to collect and analyze data. The third section explores three themes that emerged from the empirical data. The role of the convener, the development of a two-tiered program structure, and the coupling of complementary resources were collaborative techniques used within the VCZM Program to protect and enhance environmental resources. The chapter concludes by linking collaboration to the broader context of environmental management and political culture.

SETTING

The Coastal Zone Management Act of 1972 paved the way for coastal states to develop management programs focusing on protecting, restoring, and enhancing the nation's coastal resources while managing development. Faced with increasing challenges resulting from continued growth and competing demands on land use, Virginia developed a Coastal Zone Management Program in 1986. A multiorganizational arrangement of government and nongovernmental organizations worked together to protect, restore, and strengthen Virginia's coastal resources by developing coastal management practices that integrated economic and ecological sustainability.

The VCZM Program worked with federal agencies, state agencies, local governments, and nongovernmental organizations to implement the Coastal Zone Management Act by administering state laws and regulations to protect and manage Virginia's coastal resources. State agency involvement in the VCZM Program was specified by the governor of Virginia through executive order. The order designated a group of agencies responsible for implementing the program's enforceable policies and a group of agencies responsible for assisting with the VCZM Program (see Kaine 2006). In addition, the program's goals and objectives were outlined. While the impetus for the VCZM Program came from a federal mandate, a decentralized approach was used to develop and implement resource policies throughout Virginia's coastal areas. Unlike other states, Virginia did not have a central environmental planning division. Despite the designation of Virginia's Department of Environmental Quality (DEQ) as the lead agency, there was not a command and control structure used to implement the VCZM Program and the executive order did not specify how or to what extent organizations should work together during program implementation.

This arrangement was governed by a Coastal Policy Team (CPT) comprised of representatives from the state agencies involved in protecting or enhancing Virginia's eight coastal areas (Office of Ocean and Coastal Resource Management [OCRM] 2004).[2] The CPT developed policies, allocated resources, and prioritized funding streams to help implement this program. Local government representatives were linked to the VCZM Program through planning district commissions (PDCs) that fostered relationships across local government jurisdictions, created channels to pass information, and enhanced opportunities to pool resources (OCRM 2004). The VCZM Program was administered by six state employees affiliated with the DEQ's Division of Environmental Enhancement.

Projects affiliated with the VCZM Program were funded in part with grant money provided from the Office of Ocean and Coastal Resource Management within the National Oceanic and Atmospheric Administration (NOAA). States were required to have a federally approved coastal program and assess specific coastal management priority areas in order to receive federal money. Grant funds were administered by the VCZM Program staff based on priorities determined by the CPT. The money was used to maintain ongoing programs, support a large program identified as a primary focal area, and initiate smaller projects.

In 2002, the VCZM Program developed the Virginia Seaside Heritage Program and focused much of its resources and expertise on the Eastern Shore of Virginia's coast (VCZMP 2007). A peninsula with the Chesapeake Bay to its west and the Atlantic Ocean to its east, the Eastern Shore is a system of barrier islands, bays, and salt marshes that run along the coast of Virginia and Maryland. The Virginia Seaside Heritage Program focused on the portion of the peninsula adjacent to the Atlantic Ocean in Virginia. Fifteen organizations partnered with the VCZM Program to restore coastal habitats, replenish aquatic resources, promote sustainable economic activities, and develop coastal management policies on the seaside of Virginia's Eastern Shore (VCZMP 2007); these organizations are listed in table 9.1.

The Virginia Seaside Heritage Program became a main focal area for the VCZM Program in part because it was an area often overshadowed by the problems associated with the Chesapeake Bay—a body of water between mainland Virginia and the peninsula of the Eastern Shore. A common

Table 9.1. Network of Organizations in the Virginia Seaside Heritage Program

Organizational Type	Specific Organizations in the Network
Federal Agencies	U.S. Fish and Wildlife Service
Virginia State Agencies/Programs	Department of Environmental Quality Coastal Zone Management Program Marine Resources Commission Department of Conservation & Recreation Department of Game & Inland Fisheries
Local Government	Accomack County Northampton County Accomack-Northampton PDC
Nongovernmental Organizations	The Nature Conservancy Eastern Shorekeeper Southeast Expeditions Cherrystone Aquafarms
Academic Institutions	College of William & Mary Institute of Marine Science Center for Conservation Biology University of Virginia

theme among interviewees was that the seaside of the Eastern Shore was forgotten even though it was a critically important environmental area that faced severe development pressures compounded by economic stress. An interviewee described the area as "one of the world's most important biospheres." The seaside's aquatic resources, such as oysters, scallops, shellfish, and shorebirds, declined dramatically due to overharvesting, disease, and habitat loss (VCZMP 2007). Its resources are valuable from scientific, conservation, economic, and recreational perspectives. Grounds for crisis arose because funding to protect and preserve these resources was nonexistent. Much of the funding that may have been available from the state and federal levels of government was directed toward water quality issues associated with the Chesapeake Bay. This lack of funding and attention for the seaside of Virginia's Eastern Shore was in part a result of geography. Citizens and government representatives were more likely to focus on issues in closer proximity to the mainland of Virginia.

In this environmental landscape, finding solutions to counteract the depletion of natural resources and destruction of habitats was complicated by the magnitude of the problem, the inabilities of a single organization to obtain the physical and financial resources needed to resolve the problem, the number of organizations involved, and constant changes to the landscape. Projects encompassed highly varied tasks that involved different stakeholders, raised different issues, and focused on different goals. Some of the projects within the Virginia Seaside Heritage Program included shellfish restoration, shore bird habitat protection, the control of invasive plant species, and the development of ecotourism and aquaculture industries. Of the participants interviewed, more than two thirds indicated that the objectives of the program were highly complex. Many attributed this complexity to the nature and scale in which the program was trying to resolve environmental issues on the seaside. An example of this complexity could be seen when looking at a technique employees within the Division of Natural Heritage developed to map phragmites, an invasive plant species, through the use of low elevation flights with helicopters and global positioning systems. Through the use of this technique, invasive plants were located with great accuracy and in a fraction of the time required by previous efforts.

The interconnected nature of environmental resources in this landscape also contributed to the complexity of the situation. Direct and indirect relationships between predators, shore birds, invasive plants, water quality, sea grasses, conservation easements, walking trails, and the development of ecotourism and aquaculture industries needed to be understood. For example, bird habitats could be best protected by tracking predatory animals, purchasing undeveloped land, and controlling invasive plant species. Promoting ecotourism, creating trails for bird watching, and using public volunteers to plant sea grasses helped citizens understand the importance of protecting the undeveloped land around this biosphere. Through this understanding, citizens were more likely to support politically environmental planning and zoning initiatives at the local government level. It was important for partners to understand the cumulative impact of these interconnected tasks to ensure their efforts were pulled together in productive ways to obtain the desired outcome. As a result, organizations within the Virginia Seaside Heritage Program used collaborative interactions to establish an "ecosystem mentality" when focusing on land management and habitat restoration.

The Eastern Shore is comprised of farming communities where landowners faced tremendous development pressures. The VCZM Program used a two-pronged approach to curb pressures to develop large amounts of this landscape. First, much emphasis was placed on developing sustainable industries, such as ecotourism and aquaculture, which depended on the protection of coastal resource habitats. By protecting the area, these industries had opportunities to grow and create additional revenue. Second, a subset of organizations implementing the Virginia Seaside Heritage Program also worked together to purchase land on the Eastern Shore and put conservation easements in place to prevent future development.

METHODOLOGY

A single case study design was used for this research. The VCZM Program was selected as the setting through criterion purposeful sampling. Case selection was based on the following criteria: (1) an environmental program was implemented in a southern state with a traditionalistic political culture; (2) organizations associated with this program frequently interacted; and (3) the multiorganizational arrangement included various government agencies and nongovernmental organizations. This research focused on the interactions between organizations implementing the Virginia Seaside Heritage Program.

Data were collected through thirty-four semistructured interviews and a review of eight organizational documents. Information was gathered through standardized, open-ended interview questions. Interviewees were selected through snowball sampling in order to identify information rich participants involved in implementing the Virginia Seaside Heritage Program. Interviews began with members of the Coastal Policy Team. Organizational documents were reviewed to gather additional data on collaborative practices utilized during program implementation. These documents included memoranda of understanding between organizations, meeting minutes, memos, and evaluations. Data collection was conducted during April and May 2008.

A coding scheme was used to organize textual data gathered from each interview and document into themes. Data collected from interviews and organizational documents were compared. Quoted material from participants was incorporated into the discussion of themes to retain the information rich detail provided during interviews. Themes identified in this chapter were based on the elements most emphasized by interview participants and in organizational documents.

THEMES OF THE CASE STUDY

Partnering organizations worked together to manage emerging environmental issues on Virginia's Eastern Shore. Three themes emerged from this empirical research and suggest that environmental management occurred through a collaborative approach despite the traditionalistic political culture of Virginia. The convener of the multiorganizational arrangement, a two-tiered program structure, and the coupling of complementary resources are collaborative techniques used during implementation of the Virginia Seaside Heritage Program. Although this research was conducted in a single setting, environmental managers may be able to use these collaborative practices in other areas to address wicked environmental problems.

The Role of a Convener

The use of a convener was a collaborative technique mentioned frequently throughout the interviews. As suggested in the literature (see, for example, Bryson, Crosby, and Stone 2006; McNamara, Leavitt, and Morris 2010), the VCZM Program served as a mechanism to encourage interactions between organizations and establish the collective arrangement. The importance of identifying a mechanism to bring organizations together was conveyed by an interviewee: "The key [wa]s having something to bring these organizations together. Otherwise they w[ould] work together where it benefit[ed] them." Each organization involved in implementing the Virginia Seaside Heritage Program was individually busy and involved in many other projects.

With differing missions and so many organizations involved, the convening role of the VCZM Program was especially important in helping organizations recognize when they had complementary interests that could be better served by working together rather than alone. Organizations committed to the Virginia Seaside Heritage Program in part because individual missions advanced through the collective group. An interviewee acknowledged that individual organizations benefit by working together: "Beyond the funding opportunities, it g[ave] us the opportunity to do work that we otherwise would not be able to do and achieve a part of our mission that would otherwise not be possible." The VCZM Program staff recognized the commitments of individual organizations and worked to balance them with those of the collective arrangement. Collaboration occurred between partners implementing the Virginia Seaside Heritage Program because they could resolve complex environmental problems without diminishing commitments to individual organizations. The VCZM Program helped administrators see the benefits in working together and identified areas where individual organizational interests intersected. Some of the literature acknowledges that support for a multiorganizational arrangement is gained by serving the interests of individual organizations (see, for example, Keast et al. 2004; Mandell and Steelman 2003; Mullin and Daley 2009; Thomson and Perry 2006).

The VCZM Program staff gave much thought to identifying organizations whose specialized interests could be furthered through a collective effort. These organizations were invited to the table because their mission areas were tangentially related and there was an identified need for a particular expertise; collaboration between organizations occurred in a specialized way. The culmination of specialized expertise among the collective group needed to generate the capacity to address the varied environmental issues on the Eastern Shore. Duplication of expertise could result in inefficiencies or turf issues, and a lack of expertise could prevent an issue from being addressed fully. It was the VCZM Program staff who ensured a proper mixture of specialties was achieved. The role of the VCZM Program staff in aligning organizational specializations was described by an interviewee: "It's like being a conductor of a symphony. You have your different instruments and you know what their specialties are. So you figure out the right time to bring them in and hopefully it comes together in one nice piece of music." In this sense, collaborative interactions were purposive to the extent that the convener invited organizations to work together based on the specializations needed to carry out the program's objectives.

For example, the Virginia Institute of Marine Science and the Virginia Marine Resources Commission were primarily involved in oyster restoration. The Nature Conservancy and the Center for Conservation Biology were primarily involved in avian research. On the other hand, the Department of Conservation and Recreation was focused on phragmites control and providing opportunities for ecotourism. While these only represent a few examples of organizational specializations within the Virginia Seaside Heritage Program, a broader look revealed that each organization had a niche within the larger group. An interviewee explained, "certain groups [we]re involved in specific projects depending on their expertise." The multiorganizational arrangement was stabilized through repeated interactions among particular subgroups which also fostered longstanding relationships. The VCZM Program staff gave considerable thought to identifying the specializations needed to implement the Virginia Seaside Heritage Program prior to bringing the organizations together.

Fiorino (2006) suggested that conflicts can create obstacles within the current regulatory system. By utilizing a convener to bring organizations together in a specialized way, a collaborative approach to environmental management may remove these obstacles by minimizing conflict among partners. The allocation of grant money and determination of project involvement was based on the need for a particular expertise. Conflict was minimized because all organizations were involved in the decision making process to allocate funds.

Funding priorities were established through consensus. On an annual basis, the VCZM Program solicited input from all organizations involved in the Virginia Seaside Heritage Program to identify potential projects. Input was captured in a spreadsheet and distributed to all members of the CPT. They met each year to discuss the proposed projects and determine funding allocation. It was mentioned during an interview that there were projects the group supported easily. Funding discussions began with collective support for these projects which the interviewee described as the "lowest hanging fruit." Group discussions proceeded gradually toward projects where consensus would be more difficult to obtain. Through these discussions, partners developed a better understanding of one another. An interviewee explained, "everybody underst[ood] each other's work so completely that they realize[d] that none of their projects [were] more important than any others. And they [we]re looking for opportunities to find ways to help others."

The use of collaboration in environmental management may alleviate problems that plague the command and control approach and ultimately obtain a better outcome for the resources managers are trying to protect. Hierarchy, division of labor, and uniform rules are the cornerstones of a command and control approach (Fiorino 2006). The problems with this approach are that it assumes environmental issues can be compartmentalized while behaviors are prescribed through a standardized set of rules regardless of the situation at hand. It seems likely that a command and control approach would have little, if any, success at resolving wicked problems because their diverse and interdependent nature would go unrealized. The power of the Virginia Seaside Heritage Program came from working together, and partners reached across organizational boundaries to seize these opportunities. Partnerships were described during an interview in the following way: "It [was] an opportunity to be successful in a way that would be impossible otherwise. It create[d] opportunities to work with other agencies in a way where the sum of the whole [wa]s greater than the sum of the parts." Collaboration offered the best protection for environmental resources on the seaside of the Eastern Shore because the environmental landscape was addressed holistically, the complexities of this particular setting were recognized, and organizations with desired specializations were invited to become part of the resolution.

Two-tiered Program Structure

Another theme prevalent from the empirical data was the use of a two-tiered program structure to manage the organizations implementing the Virginia Seaside Heritage Program. While establishing linkages within the multiorganizational arrangement, two types of horizontal structures helped to develop and sustain collaborative interactions. The CPT was one type of horizontal structure comprised of representatives from state and local governments; each of Virginia's natural resource agencies were represented within this group.[3] A governing body provided a forum to bring organizations together. Participants made programmatic decisions to guide the overall direction of the VCZM Program and distribute grant funding. Many interviewees suggested that partnering organizations worked well together because the CPT provided the structure necessary to do so.

Since distinct legal authorities guided each agency, no organization had the authority to tell another organization what to do. Therefore, decisions were made collectively based on consensus. This process was described by an interview participant: "Decision making [wa]s a collegial process. There [we]re a lot of prioritizations to be made. It [wa]s an open, roundtable discussion. And we tr[ied] to come to consensus on what the priorities w[ould] be." Much like the literature suggests (see, for example, Agranoff 2006; Mandell and Steelman 2003), consensus and compromise were important parts of the process. It was common for interview participants to describe the process as an open discussion among equal stakeholders.

The executive steering committee was a second type of horizontal structure used to manage this collaborative arrangement. This group was comprised of operational personnel with field level expertise and responsibilities for managing projects on the Eastern Shore. During interviews, CPT members recognized frequently the experience personnel at the operational levels had in studying these ecosystems and understanding project operations. Many of the executive steering committee members worked projects on the Eastern Shore for the last twenty to thirty years.

The development and use of a two-tiered programmatic structure across organizational boundaries facilitated the involvement of resource administrators and operational project leaders from each state agency. Representatives on the CPT typically supervised the project leaders on the executive steering committee. Through these linkages, resource administrators and operational personnel became aware of the expertise within their own organizations and in other organizations. While the creation of a new program structure is consistent with the literature (see, for example, Mattessich, Murray-Close, and Monsey 2001) and other research has explored the development of a subcommittee structure to address complex environmental problems (see McNamara, Leavitt, and Morris 2010), this research suggests that management of this arrangement benefited in two ways from the horizontal linkages made at more than one organizational level.

First, high levels of trust were evident at each level within this program structure. Trust played an important role in creating and sustaining collaborative interactions as participants indicated that they felt comfortable with other members of the group. An interviewee conveyed the comfort level among partners in the following way: "Trust [wa]s built through successful accomplishment of various projects that we work[ed] on and positive reinforcement." As a result of this trust, members relied on one another regardless of organizational affiliation. This reliance was especially important for the ecosystem approach used to manage projects on the Eastern Shore because partners could not individually achieve the same goals. Reliance among researchers was explained in an interview: "In order to relate projects and do ecosystem wide research, you ha[d] to believe that other projects ha[d] value and that the research [wa]s trustworthy and that the people doing it kn[e]w what they [we]re doing. . . . [T]rust [wa]s vital to this system-wide approach." These findings suggest that organizations involved in implementing the Virginia Seaside Heritage Program employed the "ethic of collaboration" discussed by Thomson and Perry (2006, 25). Interviewees indicated that they believed their partners were committed to the collective arrangement and would work in good faith with other organizations.

According to Elazar (1984), social relationships are important to people in a state that has a traditionalistic political culture like Virginia. Throughout interviews, it was evident that personal ties bound this network together. High levels of trust created opportunities for people to become involved on personal levels. An interviewee discussed this personal involvement:"After years of working with one another, you [we]re no longer just working with an organizational face, but with a specific individual. The partnership evolve[d] from an organizational relationship into a more personal relationship. You kn[e]w who you need[ed] to call about certain things." Many of the players implementing the Virginia Seaside Heritage Program started working together almost twenty years ago to protect the mid-Atlantic migration corridor—a piece of property on the southern tip of the Eastern Shore which is an important stopover for migratory songbirds traveling from South and Central America to Canada. Twenty years later, the organizations and people representing these organizations still interacted in various ways, sometimes outside the boundaries of the VCZM Program. Interactions were described in the following manner: "These are overlapping organizations and groups that work[ed] together for different reasons. So [the Virginia Seaside Heritage Program] is one thing that pull[ed] them together but it [wa]s not the only thing that pull[ed] certain people to the tables." The organizations involved in implementing the Virginia Seaside Heritage Program had a deep understanding of their partners because they worked together in different capacities and did so for a number of years.

Second, horizontal linkages between organizations facilitated open and frequent communication among partners. Regular meetings among individuals involved in implementing the Virginia Seaside Heritage Program and routine communication among personnel working at the operational level enhanced collaborative interactions. It is through regular meetings of the CPT that partners discussed the direction of the Virginia Seaside Heritage Program, identified what their organization could provide to the collective group, and learned more about the other organizations involved in the program. The personal relationships that facilitated high levels of trust also created opportunities for personnel to communicate in the course of their daily operations. An interviewee described daily communications among organizations implementing the Virginia Seaside Heritage Program: "There [wa]s so much routine contact here that when it c[ame] time for all the partners to come together the only hard part [wa]s figuring out a date." As organizations enhanced their understanding of one another, they shared information and looked for opportunities to help one another. Since organizations focused on projects that addressed one piece of the larger ecosystem, a willingness to share information allowed them to become more knowledgeable in areas that addressed interrelated pieces of the ecosystem.

Through the use of a two-tiered structure, the VCZM Program built on relationships that already existed and did so while recognizing the importance placed on social relationships within a traditionalistic political culture. An emphasis on social relationships helped facilitate the use of a collaborative approach to environmental management in this setting. It was suggested in the literature that collaborative relationships may be particularly difficult for public administrators to sustain, as conventional bureaucratic systems emphasizing "stovepipe" specializations, hierarchical structures, and formal governance mechanisms do not inherently accommodate shared power and joint decision making (Keast, Brown, and Mandell 2007). In a collaborative arrangement, participants must have discretion to negotiate rules and make organizational decisions based on the evolution of group deliberation (Mattessich, Murray-Close, and Monsey 2001). Collaboration was possible in implementing the Virginia Seaside Heritage Program because participants leveraged informal relationships within both tiers of the programmatic structure to guide operations. According to Elazar (1984), informal interpersonal relationships provide the foundation for a traditionalistic political culture. On the surface, participants seemed largely removed from the bureaucracy typically associated with government agencies. However, more detailed conversations with interviewees revealed that these relationships were balanced by more conventional bureaucratic mechanisms. Participants may have experienced increased levels of independence and discretion, but this was based on support from leaders behind the scenes.

A balance between informal relationships and a traditional bureaucratic system was created by distinguishing between programmatic decisions and policy decisions. Informal relationships were utilized by the CPT when making programmatic decisions. However, a traditional form of bureaucracy was upheld throughout the policy decision-making process. Within the Virginia Seaside Heritage Program, local governments and federal/state agencies represented the public sector. These organizations had mandated responsibilities outlined by specific legal authorities; policy decisions ultimately occurred through the political process. In situations where program participants desired policy changes that occurred outside their programmatic boundaries, interviewees emphasized that they could influence the political process by communicating research findings through the appropriate channels. Collaboration was used at the programmatic level during implementation of the Virginia Seaside Heritage Program, but this approach was balanced with a more traditional form of bureaucracy through recognition of legal authorities and the political process.

Resource Allocation

A collaborative approach to environmental management increased abilities to pool resources; this theme was prevalent throughout the interviews. The pooling of resources enhanced the power of this collaborative arrangement. Due to the structure of the grant process, there were opportunities to pool resources around a funding stream. An interview participant explained that a one-to-one match was required for much of the grant money distributed by the VCZM Program: "The match requirement len[t] itself nicely to pooling resources. Often, the money [wa]s matched with time and personnel." It was through the matching requirement that organizations identified opportunities to leverage resources.

In addition, resources were pooled because the goals of individual organizations aligned with the projects of the collective arrangement. In some instances, organizations leveraged resources because their individual projects aligned with the goals of the Virginia Seaside Heritage Program. Several interviewees described this process as "piggybacking." Individual organizations benefited from additional funds while helping to complete projects for the VCZM Program. Other times, resources were pooled as organizations accessed funds that were unavailable to others. These funds were often applied to projects that aligned with the needs of the Virginia Seaside Heritage Program. For example, organizations pooled various funding sources in order to purchase land on the southern tip of the Eastern Shore. It was especially important to protect property in this area because development would disturb the stopover for migratory birds traveling between Canada and South and Central America. The involvement of several organizations was necessary because no one organization had enough money to purchase the land outright. In these situations, The Nature Conservancy often purchased the land initially because it was able to allocate funding in a timely manner. The Virginia Department of Conservation and Recreation and the U.S. Fish and Wildlife Service worked together to repurchase the land from The Nature Conservancy and manage it in perpetuity. All three organizations owned and managed land on the southern tip of the Eastern Shore. The importance of working together was expressed by an interviewee: "When you are faced with small pots of money, the only way to get anything done is through partnerships and leveraging people's efforts."

Nonfinancial resources were pooled in specialized ways. This theme is supported by other research (see, for example, McNamara, Leavitt, and Morris 2010). Interviewees frequently mentioned that organizations combined specialized areas of expertise in order for the collective group to achieve the goals of the Virginia Seaside Heritage Program. For example, The Nature Conservancy had access to volunteers and experience with land acquisition, the College of William and Mary had expertise in bird habitat management, the Eastern Shorekeeper monitored environmentally restored areas, the University of Virginia negotiated environmental management priorities among partners, and the VCZM Program staff members convened the arrangement and managed grant funding. These represent a few examples of the ways nonfinancial resources were pooled.

Regardless of the ways financial or nonfinancial resources were pooled, two influences from the state's traditionalistic culture were prevalent. First, resources were pooled in ways that aligned with the mandated responsibilities in each organization's legal authorities. Once the CPT decided to fund a particular project, partnering organizations carried out specific tasks that aligned with each organization's missions and goals. Alignment between tasks and organizational mission was conveyed by a resource administrator: "We ha[d] legal responsibilities that we ha[d] to [consider] to manage certain resources. Beyond that there [wa]s a mission orientation towards managing those resources." The building of an observation platform in the town of Willis Wharf provided a good example of this point. The Virginia Department of Game and Inland Fisheries (DGIF) received grant funding to design and build the observation platform. The act of designing the platform aligned nicely with the organization's mission to promote recreational opportunities for Virginians to enjoy wildlife. However, DGIF could not build the platform because it was not authorized legally to build structures on another organization's property. Since the property was owned by Northampton County, DGIF subcontracted with the county to build the observation platform. With zoning expertise, the county was better suited to build the structure.

Second, resource administrators retained control over the ways in which resources were committed to the collective group. An interviewee explained the applicability of pooled resources to the Virginia Seaside Heritage Program: "They [we]re pooled to the extent that everybody contribute[d]. They [we]re not pooled to the extent that you donate[d] a certain amount of time and somebody else decide[d] how that time [wa]s spent." This control generated a mechanism to ensure resources were used in ways that aligned with the mission and goals of individual organizations. A resource administrator explained, "[if] I [was] going to commit my time or my staff's time to [a project], it [wa]s more a matter of here [wa]s a specific task that [wa]s within our mission and my responsibility."

Collaboration was used to pool resources in specialized ways, but this was accomplished within the boundaries of a traditionalistic political culture. Informal personal relationships among partnering organizations were utilized to establish linkages for resource sharing. However, decisions were made by the administrators that controlled the resources. Through the use of a control mechanism, resources were allocated through a top-down system that ensured alignment with the missions and responsibilities of individual organizations. Personal relationships, a lack of citizen involvement, and hierarchical control are typical of a traditionalistic political culture.

CONCLUSION

The protection and management of environmental landscapes will only get more difficult as government organizations continue to face fiscal stresses, resource limitations, and personnel shortages. Multiorganizational arrangements will become increasingly prevalent as resolutions to wicked environmental problems are seemingly beyond the capabilities of individual organizations. The command and control approach to environmental management can no longer be used as a one-size-fits-all resolution to complex environmental problems. The transcendence of hierarchical structures in some multiorganizational arrangements demands a shift in environmental management.

The role of the convener, the development of a two-tiered program structure, and the coupling of complementary resources were collaborative techniques used to address the complex environmental problems on the seaside of Virginia's Eastern Shore. The VCZM Program staff understood the specialized expertise held by each organization and invited them to work together through an intersection of complementary interests. A two-tiered program structure established horizontal linkages while facilitating the involvement of resource administrators and operational personnel from each organization. Social relationships and routine communications lead to high levels of trust and understanding among partners. Coupling complementary resources increased the capacity of the collective arrangement. Goals were achieved because each organization contributed in a unique way to the implementation of the program. This environment was best protected through a collaborative approach that relied on consensus, horizontal linkages, and the leveraging of resources.

The empirical data gathered in this case study research highlights three important findings about collaboration. First, postmodern environmental management was embraced as government employees transcended bureaucratic structures to establish horizontal linkages with partnering organizations. The distribution of interview data was especially noteworthy given that the public administration literature typically associates government organizations with highly centralized and hierarchical structures; government organizations represented 60 percent of the organizations involved in this study. Of the government employees participating in this study, almost 90 percent perceived interactions to occur at a collaborative level. Instead of command and control environmental management, the convener of this arrangement developed and sustained horizontal linkages in innovative ways. Vertical linkages within individual organizations seemed less important to the collective group than the horizontal linkages that developed between organizations.

Second, certain themes were emphasized in interviews and documents more frequently than others. The presence of a convener, the development of a new program structure, and the coupling of resources permeated interview discussions. Since participants perceived interactions within their sphere to operate in a collaborative manner, prevalence of these themes suggest that they are important to consider when moving beyond command and control environmental management practices.

Third, findings from this study support other research suggesting that southern states embrace policy innovation (see Breaux et al. 2002; Breaux, Morris, and Travis 2007). The use of a collaborative management approach is innovative when utilized within the boundaries of a traditionalistic political culture. While a command and control approach is expected in southern states, participants in this setting moved beyond hierarchical confines to embrace the unexpected. A collaborative approach is possible in southern states when balanced with the traditionalistic political culture.

The VCZM Program convened a multiorganizational arrangement by inviting organizations to work together to better serve their individual goals while furthering the efforts of the collective group. By looking at the seaside holistically and recognizing interdependencies among environmental resources, the VCZM Program moved beyond the command and control approach to minimize problems associated with compartmentalization and one size fits all resolutions. A two-tiered program structure facilitated high levels of trust and horizontal linkages while leveraging the social relationships valued in a traditionalistic political culture. Participants reached across organizational boundaries while recognizing the importance of traditional patterns of responsibility, legal authorities, and the political decision-making process. Resources were pooled to help the collective achieve goals that would otherwise not be possible. Top-down decisions were made by resource administrators to control allocation and ensure alignment with organizational missions. An acknowledgement of tradition brought collaborative action within the realm of possibility. Environmental innovation occurred when the employment of collaborative techniques were balanced with political culture.

A new order for environmental management is on the horizon. The regularity of underfunded mandates will continue to challenge the command and control approach while supporting the development of multi-organizational linkages. The role of the federal government may become less influential as state, local, nonprofit, and private organizations learn to rely on one another. Findings from this research support Fiorino's (2006) suggestion that it is time to employ a new approach to environmental regulation. While the use of a command and control approach on the Eastern Shore would have likely resulted in a failure to resolve its "wicked" problems, it would be inappropriate to suggest that a collaborative approach be employed universally. Participants stressed that collaborative interactions required great amounts of time and resources to sustain. Relationships between partners involved in implementing the Virginia Seaside Heritage Program were enhanced by twenty years of working together. As found in other environmental settings (see Imperial 2005; Weber 2009), the long-standing relationships that enhanced the application of collaborative environmental management practices in this study did not develop quickly or easily. Perhaps it would be most beneficial for environmental managers to understand and employ various approaches based on differing levels of interdependencies between partnering organizations. This variety would give environmental managers options for selecting an approach that best fit a particular environmental situation or problem.

This case study indicates that the use of collaboration is not antithetical to traditionalistic political culture; rather, it appears that the networks of personal relationships so prevalent in traditionalistic cultures can be harnessed to create vibrant collaborative interactions. In the Virginia Seaside Heritage Program, personal relationships forged high levels of trust and opened communication channels among partners. These same characteristics are also described as critical components of building and sustaining collaborative relationships (see, for example, Keast et al. 2004; McNamara 2008; Thomson and Perry 2006). Successful collaboration is often thought to be innovative, and southern states are not often synonymous with innovation. It may well be the case that environmental issues are particularly ripe for successful collaboration, especially those that rely heavily on volunteers to accomplish their mission. People who believe strongly in what they are doing, as is often the case in environmental organizations, may have a natural affinity for people who feel much as they do (see McNamara, Leavitt, and Morris 2010). Such affinity, coupled with a political culture that places a high degree of emphasis on personal relationships, may provide rich opportunities for successful collaboration.

While it is possible that findings from this research may be specific to this particular environmental setting, the complexity of this landscape and a lack of funding are hardly unique. Resolutions to problems arising from complex landscapes will continue to challenge environmental managers and require the involvement of multiple organizations. The collaborative practices embraced on the seaside of Virginia's Eastern Shore may help others protect and manage environmental resources. It is through collaboration that resolutions to wicked problems become possible.

NOTES

1. Complex problems are often described within the public administration literature as "wicked" (Keast et al. 2004; Harmon and Mayer 1986; Rittel and Webber 1973). According to Harmon and Mayer, wicked problems have "no solutions, only temporary and imperfect resolutions" (1986, 9).

2. Virginia's eight coastal areas were identified as the following: Accomack Northampton, Crater, Hampton Roads, Middle Peninsula, Northern Neck, Northern Virginia, George Washington, and Richmond Regional. The setting for this study was in the Accomack Northampton coastal area.

3. The following organizations are represented on the Coastal Policy Team: Virginia Department of Environmental Quality, Virginia Marine Resource Commission, Virginia Department of Conservation and Recreation, Virginia Department of Game and Inland Fisheries, Virginia Department of Health, Virginia Department of Agriculture and Consumer Services, Virginia Department of Forestry, Virginia Department of Historic Resources, Virginia Department of Transportation, Virginia Economic Development Partnership, Virginia Institute of Marine Sciences, and representatives from each of the Virginia Planning District Commissions.

REFERENCES

Agranoff, R. 2006. Inside Collaborative Networks: Ten Lessons for Public Managers. *Public Administration Review* 66 (supplement):56–65.

Breaux, D. A., C. M. Duncan, C. D. Keller, and J. C. Morris. 2002. Welfare Reform—Mississippi Style: TANF and the Search for Accountability. *Public Administration Review* 62, 1:92–103.

Breaux, D. A., J. C. Morris, and R. Travis. 2007. Explaining Welfare Sanctions and Benefits in the South: A Regional Analysis. *American Review of Politics* 28, 1:1–18.

Bryson, J. M., B. C. Crosby, and M. M. Stone. 2006. The Design and Implementation of Cross-Sector Collaborations: Propositions from the Literature. *Public Administration Review* 66 (supplement):44–55.

Crosby, B. C. 1996. Leading in a Shared-Power World. In *Handbook of Public Administration*, edited by James L. Perry. San Francisco: Jossey-Bass Publishers.

Daley, D. M. 2008. Interdisciplinary Problems and Agency Boundaries: Exploring Effective Cross-Agency Collaboration. *Journal of Public Administration Research and Theory* 19:477–93.

Elazar, D. 1984. *American Federalism: A View from the States*, 3rd ed. New York: Harper Row.

Fiorino, D. J. 2006. *The New Environmental Regulation*. Cambridge, MA: MIT Press.

Harmon, M. M., and R. T. Mayer. 1986. *Organization Theory for Public Administration*. Boston: Little, Brown and Company.

Huxham, C. 2003. Theorizing Collaboration Practice. *Public Management Review* 5, 3:401–23.

Imperial, M. T. 2005. Using Collaboration as a Governance Strategy: Lessons from Six Watershed Management Programs. *Administration & Society* 37, 3:281–320.

Kaine, T. M. 2006. *Executive Order Number Twenty-One*. Retrieved February 4, 2008, from www.deq.state.va.us/coastal/exorder.html.

Keast, R., K. Brown, and M. Mandell. 2007. Getting the Right Mix: Unpacking Integration Meanings and Strategies. *International Public Management Journal* 10, 1:9–33.

Keast, R., M. P. Mandell, K. Brown, and G. Woolcock. 2004. Network Structures: Working Differently and Changing Expectations. *Public Administration Review* 64, 3:363–71.

McGuire, M. 2006. Collaborative Public Management: Assessing What We Know and How We Know It. *Public Administration Review* 66 (supplement):33–43.

McNamara, M. W. 2008. Exploring Interactions during Multiorganizational Policy Implementation: A Case Study of the Virginia Coastal Zone Management Program. *Dissertations and Thesis Full Text* 69, 11 (UMI No. 3338107).

McNamara, M. W., W. M. Leavitt, and J. C. Morris. 2010. Multiple-Sector Partnerships and the Engagement of Citizens in Social Marketing Campaigns. *Virginia Social Sciences Journal* 45:1–20.

Mandell, M. P., and T. A. Steelman. 2003. Understanding What Can Be Accomplished through Interorganizational Innovations: The Importance of Typologies, Context, and Management Strategies. *Public Management Review* 5, 2:197–224.

Mattessich, P. W., M. Murray-Close, and B. R. Monsey. 2001. *Collaboration: What Makes It Work?* Saint Paul: Amherst H. Wilder Foundation.

Morris, J. C., and M. Burns. 1997. Rethinking the Interorganizational Environments of Public Organizations. *Southeastern Political Review* 25, 1:3–25.

Mullin, M., and D. M. Daley. n.d. Working with the State: Exploring Interagency Collaboration within a Federalist System. *Journal of Public Administration Research and Theory.* Advance Access published October 29, 2009, doi:10.1093/jopart/mup029.

Office of Ocean and Coastal Resource Management. 2004. Evaluation Findings for the Virginia Coastal Management Program: November 1999 through July 2003. Washington, DC: National Oceanic and Atmospheric Administration.

Rittel, H. W. J., and M. M. Webber. 1973. Dilemmas in a General Theory of Planning. *Policy Sciences* 4, 1:155–69.

Rosenbaum, W. A. 2005. *Environmental Politics and Policy.* Washington, DC: CQ Press.

Thomson, A. M., and J. L. Perry. 2006. Collaboration Processes: Inside the Black Box. *Public Administration Review* 66 (supplement):20–32.

Virginia Coastal Zone Management Program. 2007. Virginia Seaside Heritage Program: Goals and project highlights 2002 through 2007. Richmond, VA: Author.

Weber, E. P. 2009. Explaining Institutional Change in Tough Cases of Collaboration: "Ideas" in the Blackfoot Watershed. *Public Administration Review* 69, 2:314–27.

Wood, D. J. and B. Gray. 1991. Toward a Comprehensive Theory of Collaboration. *Journal of Applied Behavioral Science* 27, 2:139–62.

10

The Biofuel Policy in the American Southeast

How Will the Southern States Manage the Potential?

Erin Holmes

INTRODUCTION

Biomass fuels or biofuels provide energy through combustion of organic matter derived from non-fossil fuels. These fuels may come from recovery of energy from waste products, crops grown specifically for their energy value, or some combination of these. Current energy production technologies for these fuels hold promise for cleaner, cheaper, and more secure sources of energy than coal and oil as well as economic development potential for rural America. These promises for renewable and sustainable energy in the face of the rising costs of traditional energy are accompanied by a host of unanticipated policy outcomes including the potential for increased environmental degradation and the threat to food security and affordability, and leave open to question the appropriate roles for government in the development of the biofuel market. Policy questions surrounding a state's agricultural potential to produce biofuels must also address a state's ability to manage that potential through policy infrastructure and design. Very little work has been done to examine the types of policies and policy instruments in the Southeast and how that policy infrastructure can support and augment the Southeast's biofuel potential. The question of biofuel policy infrastructure is the heart of this chapter. It begins with the examination of how southern states are positioned to adjust to changing agricultural and biofuel energy markets, to move away from traditional energy sources, and to manage potential environmental impacts of biofuel production.

This chapter examines the importance of southern biofuel policies within the framework policy instrument choice and the larger environmental, energy, and agricultural policies. The potential for positive and negative policy consequences as the southern states shift to biofuel production and the relative lack of knowledge regarding existing biofuel policy instruments and the creation and evolution of that policy within the southern states makes this topic and research both important and timely.

BACKGROUND

Southern agriculture and agricultural practices have long been acknowledged as a cultural and political force by southern scholars and researchers. V. O. Key describes many of the region's social and political traditions as evolving from agricultural practices found in the "black belts" of the South or in terms of the land and its workers (Key 1949). Ira Sharkansky states that much of the South's poverty, racial tensions, and social problems are a result of its "poor-soil rural economy" that impoverished not only the citizens but the state governments as well (Sharkansky 1970). And while contemporary Americans, both southern and non-southern, like to think of the South in terms of its more cosmopolitan cities such as Atlanta, Nashville, or Raleigh, much of the South is still very dependent on agriculture and the land. According to the Bureau of Economic Analysis, agricultural production in the southeastern states account for nearly thirty-seven billion dollars of the region's gross state domestic product. The dependence of southern states on agricultural production ranges from 4 percent of state GDP in Arkansas to only 0.5 percent of the state GDP for Virginia (Bureau of Economic Analysis 2008). This agricultural dependence coupled with the increased importance of energy to the American economy creates an environment ripe for the development of biofuels and the requisite policies.

Biofuels' growth into a viable part of the nation's energy portfolio is uncertain but vital for economic development and national security. Concern for energy security in the United States has increased sharply since 2007 when President Bush announced his "Twenty in Ten" initiative. The goal of this initiative has been to reduce gasoline consumption over the next ten years through the substitution of biofuels (Bush 2007).

Less than a year into his presidency, Mr. Obama announced the development of a biofuels interagency working group charged with the development of biofuel markets, coordination of infrastructure policies, and restructuring investments in biofuels to secure the future green jobs in the country (The White House 2010). In May 2009, Energy Secretary Steven Chu announced a $785 million investment from the American Recovery and Reinvestment Act specifically targeted to develop biorefinery capacity, basic biofuels research, and ethanol research (U.S. Department of Energy 2009).

As an emerging part of the country's energy portfolio, biofuels are part of a larger class of alternative or renewable fuels that include wind, solar, geothermal, tidal energy, and fuel cells. It is ironic that as the United States Department of Energy celebrated the thirtieth anniversary of its inception during the nation's first real energy crisis, the focus of energy policy has again shifted to concern over record high traditional fuel prices and energy security (Bodman 2008). The combination of increased global consumption of petroleum and petroleum products resulting from industrialization and economic turbulence in the United States is the main impetus for the shift in energy policy focus. The United States' increasing dependence on the importation of traditional energy sources—oil, coal, and natural gas—leaves the stability of its economy exposed to threats from supply disruption and high prices. Security threats, market disruptions from natural disasters, and unstable prices of crude oil make conservation of traditional fuels desirable and the development of biomass or renewable sources economically feasible. The more recent decline in energy prices attributed to the global recession has underscored the need for stability in the energy markets. The Energy Information Administration (EIA) of the U.S. Department of Energy, along with other economic experts, predicts that the price of fuel or the consumption of traditional energy stock will return to high prices as the world climbs out of recession and the demand for energy increases. Further, energy policy experts at the EIA expect that the consumption of nontraditional biofuel energy will rise as the nation moves from the production of eight million barrels per day currently produced to ten million barrels per day by 2030 (Energy Information Administration 2008). The consumption of energy supplied by biomass sources will rise from over two quadrillion British Thermal Units (BTU) in 2005 to four quadrillion BTU by 2030, an increase in consumption of approximately 71 percent. Over the same period EIA estimates that biofuel production will nearly double from almost three quadrillion BTUs to slightly over five quadrillion BTU (Energy Information Administration 2007). Calls for the widespread use and expansion of other alternative fuels such as algal biofuels, geothermal projects, methane gas recovery, development of wind, and solar capabilities have risen sharply from both inside and outside of government (U.S. Department of Energy 2009; Pickens 2008; United States Department of Energy 2008).

New technologies are being developed and deployed that allow for the production of biofuels to move away from the traditional corn-based feedstock to less input intensive feedstock such as algae, switchgrass, and wood. Though the states in the southeastern region of the country are the states with the greatest potential crop yields for production of cellulosic ethanol, the University of Georgia was the only institution in the Southeast to receive part of the ten million dollars in grant funding announced in July of 2008 by the U.S. Department of Energy and the U.S. Department of Agriculture for the study biomass genomics and cellulosic ethanol technology (Energy

Efficiency and Renewable Energy Biomass Program 2008; Antares Group 2007). This solo award might simply be the result of a highly competitive federal grant environment or it may well be indicative of a lack of biofuel policy infrastructure, the absence of an alternative fuel vision among the southeastern states, or both.

The impact of biofuel development is recognized from all corners of the agricultural policy field. First, the American farmer clearly recognizes the potential impact of biofuels on rural America and agricultural profitability. Leading up to the 2007 Farm Bill, the United States Department of Agriculture surveyed American farmers and respondents who overwhelmingly recognized the increased importance of biofuel crop production, indicating that the development of biofuels would "help propel a major renaissance of agricultural economic prosperity" (United States Department of Agriculture 2008, 2). The American Farm Bureau (AFB), the "voice of agriculture," offers strong support for tax incentives that will offer stability and encourage research and diversity in ethanol feedstock development, particularly cellulosic ethanol. The goal of the AFB is to position the American farmer in such a way as to maximize both crop-for-food and crop-for-energy production (American Farm Bureau Federation 2009).

Nonagricultural groups recognize the importance of biofuel production on the American farm economy as well. Ethanol production from corn waste or stover could add up to nine billion dollars to the economy and as many as 76,000 permanent jobs in rural America (Common Purpose Institute 2006a). The growth in ethanol production is changing agriculture throughout the United States as the profit margins for ethanol increase with the increasing price of oil and gas (Westcott 2007). In its quick facts on Biomass Energy Crops, the Common Purpose Institute stated that if Florida's electricity producers substituted two percent of its fossil fuel for Florida grown biomass energy crops it would have an impact of $100 million on Florida's agricultural industry (Common Purpose Institute 2006a). Switchgrass production for biomass has been estimated to have the potential to raise United States' farm income by six billion dollars. The use of switchgrass and other fast growing woody feedstock in second generation biofuel technologies has the potential to significantly reduce both the production costs of biofuels as well as the environmental impacts of production and is particularly important to the southeastern states (Koonan 2006; English et al. 2006). The states in the southeastern United States are estimated to have the highest potential for the production of cellulosic ethanol from the production of woody crops and second highest potential for total production behind the major ethanol feedstock production region of the north central United States (Antares Group 2007).

Concerns for environmental stewardship encompass both farm and rural interests as well as traditional environmental interests. A joint report issued by the Natural Resources Defense Council (NRDC), the Western Resource Advocates, and the Pembina Institute, reflects the concerns of the environmental community of the country being "at an energy crossroad[s]" where the policy choices are between "ever dirtier" energy sources and those that promise "a more sustainable energy future" (Bordetsky et al. 2007, 4).

At this policy crossroads are questions of practices that can help or harm the environment. Biofuel production can enhance farming practices through the use of "bridge farming," a technique that focuses on plantings that can be used to transform less productive agricultural lands into more productive lands by building up soil nutrients (Southern States Energy Board 2005). Other environmental advantages of biomass crop plantings are under investigation in many locations. Among the positive effects that may be garnered by large-scale plantings of high cellulose crops such as switchgrass or fast growing trees include lower levels of long-term erosion, improved wildlife habitat, and the reduction of chemical run-off from farm lands (Tolbert and Downing 1995).

There are significant and very real concerns regarding the impact of large scale biofuel production on the environment and whether the production and use of these fuels will result in net reductions of significant greenhouse gases. To counter the suggestion that biofuel production will result in better farming practices, the Environmental Working Groups is concerned that the production of corn based biofuels will substantially increase soil erosion, fertilizer run-off and subsequent water pollution, and increased water loss due to corn irrigation practices (Environmental Working Group 2007). Defenders of Wildlife predict that increased corn production will result in the destruction of habitat, not the development of habitat, as farmers move currently fallow land into production (Defenders of Wildlife 2008).

Of significant concern to environmental interests are the increasing levels of carbon and sulfur dioxide emissions. Demand for electricity and increased transportation needs have been the main drivers behind the rise in carbon dioxide emissions in the United States since the 1960s (United States Public Interest Group 2007). The environmental impact of biomass and biofuels on these traditional pollution concerns is still controversial. Wald in *Scientific American* questions the utility of ethanol because of the high energy inputs necessary to grow current ethanol feedstock and corn, and to distill it (Wald 2007). The NRDC contends that the change from gas to cellulosic ethanol can reduce global warming pollutants by 88 percent (Natural Resources Defense Council 2007). The U.S. Department of Energy holds a similar view that corn ethanol, if examined over the entire life cycle of production reduces greenhouse gas emissions by 19 percent compared to gasoline (United States Department of Energy 2008).

Finally, concerns about the environmental impact of the leading fuel oxygenate, methyl tertiary butyl ether (MTBE), and its replacement with ethanol have led to concerns over the available supply of ethanol. In its survey of eighty-four U.S. ethanol producers in 2001, the California Energy Commission found that thirteen new plants were under construction with thirty-four new plants planned. Total production capacity was expected to double by 2005. However, the most interesting outcome of this survey was that the Gulf States only accounted for two of the 57 total plants in the country in 2001 and were only expected to add two more plants by 2005. As a result, the Gulf States are expected to account for less than 4 percent of the ethanol production facilities and less than 2 percent of total expected output (MacDonald, Yowell, and McCormack 2001).

American farmers and environmental groups have policy positions and interests in biofuel policy, but what about American business groups? The American business community has a very real interest in the development of a stable energy source for manufacturing and production. The National Association of Manufacturers (NAM) maintains that higher energy costs translate directly to the loss of three million jobs, cost specific industries billions of dollars in increased energy costs, and force the closure of manufacturing plants (McCoy 2007). Transportation and shipping costs are particularly sensitive to the recent volatility in energy prices. Since 1980, American manufacturers' dependence on shipping by truck has doubled in terms of truck miles traveled and increased transportation fuel costs have hit American businesses particularly hard in recent months (Federal Highway Administration 2008).

It is clear that the biofuel energy stakes are high for American businesses but how does this translate into policy or policy goals? The U.S. Chamber of Commerce supports domestic development of traditional energy sources such as oil, natural gas, and coal (Coyne 2008). It favors more drilling and a "comprehensive and balanced energy strategy" that also includes the development of clean energy from wind, solar, and geothermal sources and a self-sustaining, stable market (Institute for 21st Century Energy 2008, 6). The National Association of Manufacturers (NAM) favor energy policies that feature public/private partnerships, increased education about energy use, efficiency, and a "rationalizing of existing statutes and regulations" that will make the business environment less complicated and more productive (McCoy 2007).

How are all of these interests addressed in policy? What are the mechanisms that government uses to encourage or discourage particular energy consumption behaviors? In a study that ranked the states on the policies they adopt to promote green energy production, the Union of Concerned Scientists found that states generally used four mechanisms to promote green energy practices including renewable electricity generation standards, the establishment of dedicated funds for the development of renewable energy programs, net metering policies, and fuel and emission disclosures

(Clemmer, Paulos, and Nogee 2000). The Southern States Energy Board conducted a survey of representatives of the bioenergy industry in 2003 to gauge levels of policy awareness these industry professionals had with regards to federal, state, and local bioenergy policies. The most important finding of this research appears to be that industry officials, the most knowledgeable biomass professionals, know very little about existing government bioenergy policies. As a result, programs and policies appear to be less effective than hoped (Badger 2003).

Are there any discernable patterns to the types of biofuel policy instruments selected by the southern states and the various groups discussed above—the farmer, the environmental activist, and the business member? The next section on the theoretical framework explores the patterns in state-level biofuel policy making on policy instrument selection. It begins by outlining policy instrument theory and the relationships between interest group preferences and the realities of southern biofuel policies.

POLICY INSTRUMENTS IN THE SOUTHERN CONTEXT

The salience of the issues to the various groups discussed above demonstrates the complexity of developing viable biomass energy policy and the number of potential actors involved in the policy decisions. There is a different position or policy proposal for each of the interest groups and this chapter does not enumerate all of the policy issues necessary to develop a comprehensive, integrated biofuel policy. Good biomass energy policy development does include policy changes in agriculture, crop subsidies, planting practices, commodity markets, energy and environmental policies, interstate commerce, price supports for farmers, utility and traditional energy industries, mining, and transportation. These changes in policy get at the core of what government does, how it encourages or discourages behavior among its citizens, and what tools it uses to deliver the policy outcomes its citizens desire. There has yet to be a discussion of exactly what governments are doing to address the energy issues and what tools they choose to use. This discussion now turns to how governments do what they do, what policy instruments exist, what instruments appear to be favored by the southeastern states, and what these choices say about the South's approach to biofuels.

Policy scholars and others say that governments can do three things: tax, spend, and regulate (Kraft and Furlong 2006). While this is true, the question of how governments accomplish these three tasks is of increasing importance to public policy and administration scholars. The choices states make regarding policy instruments offer hints to the citizens' perspective on the appropriate role of government and who should benefit from or bear the burdens of governance and policy goals (Ingram, Schneider, and DeLeon 2007). As such, what do the current policy instruments selected by the southeastern states to address public problems surrounding biofuels say about each state's ability to address biofuel policy and its approach to governance in this policy arena?

Before that question can be answered, it is important to ask about the focus on the southeastern states in regards to biofuel policy. Why not focus on the upper Midwest where there is a long history of biofuel production and policy development? The reasons for this choice are clear: States of the southeast reflect several very important factors or regional characteristics of interest to biofuel policy, its production, and its distribution. First, the Southeast's global environmental impact is substantial. The amount of carbon dioxide emitted by the twelve southeastern states under review here has increased by 20 percent since 1990 compared to only a 12 percent increase for the rest of the United States (Environmental Protection Agency 2003). Since the United States is the leading emitter of carbon dioxide in the world, the southeastern United States' emissions likely surpass many other nations (Common Purpose Institute 2006b).

Second, the Southeast's diverse agricultural production, its economic dependence on agriculture, and the presence and man-made transportation infrastructure makes the potential for second generation biomass fuel production from switchgrass and woody feedstock not only feasible but lucrative in terms of actual production and economic development for rural communities. The region's long planting season permits higher biomass crop production and efficient rotation (English et al. 2006). Also, southern transportation infrastructure including interstate access, ports, and airports allows for both cost effective transportation of biomass fuels and built-in demand for product. One of the inherent problems with ethanol is that unlike petroleum products it cannot currently be shipped through pipelines. The extensive and well-developed shipping infrastructure in the Southeast will prove to be an important asset for biofuel producers.

Finally, other support for examining the southeastern states can be found in the responses that southeastern producers had regarding the priorities for the 2007 Farm Bill. Unlike their colleagues in the north central part of the nation, where alternative fuels are already big business, southeastern farmers tended to rank safe and secure food supply higher than energy crops as a goal for the 2007 Farm Bill (Lubben et al. 2006). This very traditional view of the role of agriculture indicates there is opportunity in the biofuel product market for entrepreneurial agriculture and policy actors even as there may be few mechanisms in place to guide biomass fuel production.

Before any discussion of policy instruments in the southeastern states can be addressed it is necessary to clarify what is meant here by the term "policy instrument" and its importance for public policy. A simplistic definition of "instrument" serves as a good foundation for understanding instrument theory. Merriam Webster defines instrument as a means to an end or a device used to achieve a goal (Merriam-Webster 2008). Moving from the general to the specific, a policy instrument is "a tool of public action," the goal of which is to address a public problem or "collective action" problem (Salamon 2002, 3). The concept of policy instrument has been expanded to include the government's techniques to change behavior among citizens and include

> the elements in policy design that cause agents or targets to do something they would not otherwise do with the intention of modifying behavior to solve public problems. (Schneider and Ingram 1997, 93)

Lascoumes and LeGales contend that instruments carry with them a history of use, scope, and social meaning. As stable sets of rules and behavioral frameworks, policy instruments are institutions that are used as "a means of orienting relations between political society and civil society" (Lascoumes and LeGales 2007, 7). Traditional policy analysis and public administration scholarship has left the study of policy instruments relatively untouched. The general position is that the instruments used are secondary to the process and are "a purely superficial dimension" of government action (Lascoumes and LeGales 2007, 2).

An examination of both an earlier era of American government and more recent governance trends reveals the importance of developing a more thorough understanding of policy instruments. In his examination of American Federalism, Daniel Elazar discusses early collaborative efforts of federal, state, and local governments with private entities and private citizens to secure expansion into the western territories. Instruments used during this expansion era included land grants to private and public entities, the development of educational resources for settlers and farmers in the form of the United States Agricultural Extension Service, as well as the founding of many public land grant institutions that provided formal education and resources to the nation's citizens. These government services and initiatives are early forms of government instruments that have been largely overlooked by public administration scholars and contemporary political scientists (Elazar 1984, 2001).

Another example of the use of instruments in governance and their importance can be found in the "reinvention" movement that began with the publication of Osborne and Gaebler's *Reinventing Government* (Osborne and Gaebler 1992). The main thrust of their work and those that follow is that government should shift from the traditional command and control, centralized authority tools to tools of privatization and collaborative partnerships to develop a government that "works better costs less" (Kettl and DiIulio 1995). Many of the federal and state initiatives during the last part of the twentieth century were focused on this goal, including those espoused by Vice President Al Gore and the Winter Commission's *Hard Truths* (The National Commission on the State and Local Public Service 1993; Kettl 1994). These reform movements have advocated the use of particular policy instruments with little regard to the need for public administration practitioners to have information about how the instruments will work and what the impact of their deployment will be on the government operations. This lack of knowledge forces public administrators to operate by "trial and error" and reinforces the need to understand more about policy instruments, their use, and their impacts (Blair 2002; Peters and Linder 1998; Linder and Peters 1984).

As part of the effort to understand policy instruments, many scholars have attempted to develop taxonomies or classifications of instruments. Theodore Lowi's four part classification is one of the first typologies offered by public policy scholars. His classification of regulatory, distributive, redistributive, and constituent policy types represents an important step forward for policy analysis research (Shafritz and Borick 2008; Lowi 1972; Kraft and Furlong 2006). A fundamental problem with Lowi's typology and others like it is that it is very difficult to correctly categorize many policies into the categories while still meeting the criteria fundamental to typological frameworks of categories that are mutually exclusive and exhaustive (Bailey 1994).

Second, many policy instruments are too complex for simple classification and instruments commonly change over time into a different type of policy instrument (Smith 2002; Greenberg et al. 1977). Policy shift over time can be dismissed relatively easily if one understands that most policy analysis is done using cross-sectional methods making policy shifts over-time less of an issue (Bailey 1994). To address the complexity issue, policy analysts have shifted the focus of instrument research to develop phenomenological understanding of policy instruments. This approach tries to understand the meanings, interpretations, and the social constructions that define instruments (Steinberger 1980; Schneider and Ingram 1997). Since typologies are fundamentally conceptual, this approach to the study of policy instruments should prove both valid and fruitful (Bailey 1994).

A more recent addition to policy instrument taxonomies is the work of Lester Salamon. He approaches policy instruments from a more phenomenological approach than Lowi as he discusses policy in terms of the evaluative criteria and the dimensions or characteristics of the policy. The evaluative criteria discussed by Salamon are those familiar to every student in an introductory policy analysis course, the criteria of effectiveness, efficiency, equity, and political feasibility (Kraft and Furlong 2006; Peters 2004). Of particular interest here, however, are the dimensions that

Salamon outlines or describes as his four "analytics": coerciveness, direct-ness, automaticity, and visibility (Salamon 2002). The research here will focus on the dimensions of directness and coercion. The reasons for this focus will be discussed below.

Directness is described by Salamon as the extent to which government is involved in funding, management, and the delivery of a public program (Beam and Conlan 2002). The traditional ideal type of government policy is one that is funded, administered, and delivered entirely by government agencies (Weber 1958; Salamon 2002, 1981). These are the types of pro-grams that have been assailed by the waves of contemporary reform and are considered to be inefficient and ineffective by public choice theorists and other economics-based public administration theorists as well as by many New Public Management scholars (Blair 2002; Ostrom 1989; Salamon 1981). As a result of these reforms, it should be expected that because bio-fuel policies are relatively new, they should not exhibit high levels of direct-ness, particularly in the Southeast where conservative ideology is strong.

Coercion is ubiquitous in public policy and policy instruments because it is through these instruments that governments shape citizen behavior and get "people to do things that they might not have done otherwise" (Schnei-der and Ingram 1990, 513). Schnieder and Ingram suggest that coercion is present in all policy instruments, even those that seem to present only ben-efits to recipients. Instruments that are constructed as incentives or to build citizen capacity have a dual nature of denial of benefits if behavior is not what is expected or desired (Schneider and Ingram 1993). It is this duality of nature and the implication of sanctions that makes coercion the most interesting of the dimensions suggested by Salamon and the reason for the focus in this research (Derthick 2001; Cho and Wright 2001). Additionally, as suggested above with the dimension of directness, many of the reforms during the last part of the twentieth century focused on policy reform that should have reduced coercion levels. As such, the states of the Southeast should enact policies with relatively low levels of coercion.

METHODS AND FINDINGS

The twelve southeastern states were examined based on active biofuel policies. Questions explored included the types of instruments chosen and the important dimensions of those instruments. The method used in the analysis employed the instrument choice framework of Salamon (Sal-amon 2002). Each instrument is rated as low, medium, or high in levels of directness and coercion. Levels of coercion and directness are defined by "the extent to which a tool restricts individual or group behavior as opposed to merely encouraging or discouraging it" (Salamon 2002, 25).

Low coercion instruments rely solely on voluntary cooperation. Medium coercion tools are still voluntary but the targets must pay a fee, a fine or tax to participate in the behavior or activity. To encourage particular behaviors, government may offer a direct subsidy that allows the citizen to do with that subsidy what he wishes. High coercion instruments are those that "impose formal limitations" on behavior that is problematic or needs control. The directness of an instrument is measured by the authors of *The Tools of Government* based on the financing and delivery system as well as an agency serving monopolistic functions of authorization, funding, staffing, and delivery of services.

There are fourteen instruments identified by the authors of *The Tools of Government*. Two of the fourteen were identified as high on both of the dimensions of directness and coercion: direct government and economic regulation. Direct government instruments feature government finance and delivery of services and economic regulation directs the market through the regulation of the behaviors and activities of private industry (Leman 2002).

Direct loans, government corporations, and government insurance are instruments that are defined as high in direct government activity but with moderate levels of coercion on the actors within the policy arena. These instruments include activities of direct government lending to actors and government servicing of the loans, the creation of a wholly owned government corporation that is a legally separate entity from government, and government shouldering of the burden of risk due to specific events (Feldman 2002; Stanton and Moe 2002).

The only instrument remaining that is defined as being highly directed by government is public information. Defined as low in levels of coercion, public information is the development and delivery of educational and informational services that are intended to modify citizen behaviors. Public information campaigns abound in the United States within many policy arenas including environmental policy initiatives. Initiatives for waste oil disposal, litter abatement, and fire control in the national park system are some examples (Weiss 2002).

Social regulations are instruments of command and control that are moderate in levels of direct government action while high in levels of coercion. These instruments are designed to control behavior of citizens within the policy arena for health and safety reasons. These are ubiquitous government instruments and include everything from traffic regulations to local animal control laws (May 2002).

Two instruments were identified as moderate along both policy dimensions: contracting and corrective taxes. Contracting was defined as the purchase or procurement of goods and services for government through private industry. Corrective taxes use taxes to create behavior change among the actors within a policy field (Cordes 2002; Kelman 2002).

The tax expenditure instrument was the single instrument defined as being moderate along the dimension of directness and low along the dimension of coercion. The instrument is defined as a deferral of tax obligation by government in order to encourage particular behaviors among citizens or to encourage the development of strategic industries in the marketplace (Howard 2002).

The four remaining instruments defined by Salamon are all categorized as low on the dimension of directness and moderate on the coercion dimension. These instruments are government sponsored enterprises, grants or grants-in-aid, loan guarantees, and vouchers. The government-sponsored instrument differs from the government corporation tool discussed above because it is defined as a privately owned and controlled business that is chartered by the government. Grants or grants-in-aid are quite familiar to policy scholars. These government instruments are cash gifts to identified or qualified groups that are used to encourage services or behaviors among the target group. Familiar examples include grants to college students to attend school or grants to social services which allow organizations to offer or extend services to defined groups. Loan guarantees are an indirect instrument that contrasts with direct government lending because the administration of the loans is conducted by external entities such as private banks but the government guarantees any losses that might occur due to specific events (Stanton and Moe 2002; Beam and Conlan 2002; Stanton 2002; Steuerle and Twombly 2002).

State level instrument choice data from the year 2008 were gathered from existing data collected by the United States Department of Energy, the National Renewable Energy Lab, and the Alternative Fuels and Advanced Vehicle Data Center (Alternative Fuels and Advanced Vehicles Data Center 2008). The data were subsequently categorized into the instruments outlined above. However, during this categorization process it became apparent that a category of government tool was not accounted for in Salamon's typology. This policy instrument involves actions taken by state government to encourage the use of biofuels by state agencies themselves. These internal rules or as Lowi calls them, constituent rules, set renewable fuel standards (RFS) for state governments that require a certain level of renewable fuel use by the state by a certain date (Lowi 1972). These standards, while they have an internal impact on state government operations, also have the external impact of encouraging biofuel market development by guaranteeing a certain level of fuel purchases by state governments while the general consumer market develops. Because Salamon does not recognize Lowi's constituent category, it is necessary to include these types of state policies into a new group of instruments labeled "market development." Based on the criteria for directness and coercion established by Salamon, the market development instrument is a highly direct instrument but low in coercion (Salamon 2002).

There are ninety-five individual policy instruments enacted by the southern states. North Carolina has enacted a total of seventeen biofuel policy instruments, the highest number of instruments enacted by any of the southern states. Alabama has enacted only one policy instrument. Those instruments not found within the policy portfolio of the southern states include direct government, insurance, government corporations, contracting, vouchers, loan guarantees, and government-sponsored enterprises. The grants/grants-in-aid and tax expenditure instruments are the most commonly used instruments among the southern states and only one state, Virginia, uses the direct lending instrument.

Cluster analysis was used to examine the southeastern states and their biofuel policies. Cluster analysis uses defining group characteristics to represent "homogenous" groups of states (Obinger and Wagschal 2001). The defining characteristics selected identified each state's policy instrument approach through indicators of the intervention a state is willing to make in the arena as represented by the overall number of policies, and the state's philosophy regarding the overall approach it should take as determined by the types of instruments selected (Schneider and Ingram 1990; Peters 2000; Linder and Peters 1981).

The cluster analysis is conducted using Ward's method, a hierarchical agglomerative method of cluster analysis through SPSS (Aldenderfer and Blashfield 1984). The sample size of only twelve states necessitates the use of Ward's method and makes it the most appropriate choice for analysis. This hierarchical method also does not require the identification of a number of clusters before the clustering procedure begins which is not appropriate for this research.

The results of the analysis indicate that there are three clusters in the group of twelve states. Cluster one includes the states of Alabama, Mississippi, and Arkansas. Cluster two contains the majority of the states with eight members: Florida, Louisiana, Georgia, Kentucky, South Carolina, Tennessee, Texas, and Virginia. Finally, cluster three includes only North Carolina. It is necessary to confirm that the cluster solution is appropriate. This is done through two methods. First, the "elbow" point that represents the point which the within-group difference becomes dramatically different was identified (Obinger and Wagschal 2001; Castles and Obinger 2008; Mazzocchi 2008). Second, the Ward method was followed by a k-means clustering process that both verifies the solution and identifies important variables to the solution. Results of both methods confirmed the initial solution (Castles and Obinger 2008). The k-means clustering indicates that the total numbers of policies, corrective taxes, direct loans, economic regulation, and grants/grants-in-aid are the most important variables that define group differences the clustering solution.

As discussed above, the cluster solution results in a single state, North Carolina, in cluster three. While outliers in ordinary least squares regression may be a cause for concern in model creation, in cluster analysis, outliers serve to identify uniqueness in the data that provides context and are worth investigating further. Outliers in cluster analysis do not represent serious flaws in the data (Aldenderfer and Blashfield 1984).

An examination of each cluster presents different behavioral pictures among the southeastern states as well. The lone case of North Carolina stands starkly against the other two clusters because of the number of biofuel policy instruments it has enacted. The state alone accounts for 18 percent of all biofuel polices in the Southeast. However, the pattern of instrument choice indicates the repeated use of a set of instruments, with five instruments used most often: corrective taxes, economic regulation, grants/grants-in-aid, market development and tax expenditures. North Carolina accounts for almost 30 percent of the total grants/grants-in-aid and nearly 31 percent of the total economic regulation policies in the Southeast. The literature on instrument preferences does discuss the reasons for repeated choices indicated by the pattern of North Carolina. It could indicate that the instruments work very well for the state and its policy makers return to them as tried and true policy approaches. Or, more interestingly, it could also indicate an ideological bias for the choice of instruments made by North Carolina (Linder and Peters 1981). The limitation of sample size makes exploring these questions not possible here.

North Carolina's instrument choice is the most coercive biofuel policy portfolio of all three groups. The average level of the coercion dimension of the instruments chosen by North Carolina is the highest of the clusters at 1.94 (on a scale of 1 to 3 with 1 being low coercion and 3 being high coercion). This rate makes it 13 percent higher than the average and 20 percent higher than cluster one with the lowest average on the coercion dimension. North Carolina has the lowest average on the direct government dimension, however. It is nearly 10 percent lower than the average for the Southeast and is 13 percent lower than cluster one, the group with this highest average direct government dimension.

The largest of the clusters includes eight of the twelve southeastern states: Florida, Louisiana, Georgia, Kentucky, South Carolina, Tennessee, Texas, and Virginia. These states account for 73 percent of the total biofuel policy instruments in the Southeast. This cluster includes a total of seventy instruments, or an average of ten biofuel policies per state. Tax expenditures make up the bulk of its portfolio, accounting for 21 percent of the sixty-eight instruments. This cluster accounts for 82 percent of the total tax expenditures in the Southeast. Public information is the next most frequently used policy instrument, and it accounts for 19 percent of the cluster total and

87 percent of the public information instruments in the Southeast. Unlike North Carolina and the states in cluster one, cluster two has the most variety of policy instruments in its portfolio with eight different instruments chosen by its member states. While its policy portfolio indicates clear favorites, such as public information, tax expenditures, and corrective taxes, it exhibits a stronger tendency to experiment with instruments than either of the other clusters. At the same time the states of this cluster experiment with policy instruments, they also occupy the middle ground on both the directness and coercion dimensions for nearly all the southeastern states.

Cluster one consists of three states, Alabama, Arkansas, and Mississippi, which account for eight of the nine instruments enacted by the southeastern states. While North Carolina stands out for the number of biofuel policies it has enacted, these states stand out for the number they have not enacted. Alabama has enacted only one policy pertaining to biofuels, leaving Mississippi and Arkansas to account for the remainder of the policy instruments. The instruments in this cluster's portfolio include corrective taxes, grants/grants-in-aid, public information, market development, and social regulation. Even though it has the smallest number of instruments in the Southeast, the portfolio is both the most direct and the least coercive of the three clusters.

IMPLICATIONS

What does it all mean? First, the South is not as homogenous, at least as it addresses its biofuel policies, as some contemporary scholars would have others believe. It is truly representative of the diversity that was found by Key (1949) and other southern scholars that was somehow lost in the attempts to discuss exceptionalism. There is no single "best" policy approach when it comes to southern biofuel policy and certainly the conservatism that is so uniformly applied to the South does not bear fruit in the biofuel policy arena. The policies of North Carolina bare very little resemblance to those of the rest of the southern states and the policies of the other groups do not indicate any type of policy homogeneity. To understand policy making in general and biofuel policies in particular, one must break through the stereotypical picture of the homogenous South and understand each state in terms of its unique ability to adapt and change its policies to its circumstances.

The picture of the biofuel policies in the South appears to be diverse, confusing, and not at all coordinated, connected, or uniform. This appearance of diversity is the most important implication of the research. This also reveals that while in many ways the behaviors of southern policy makers, politicians, and citizens are very similar, they are much more different than common perceptions hold. If the institutions of the southern policy environment were homogenous, one would expect to see biofuel systems that are more alike than they are different. In a situation of true homogeneity, there would have been only a single group or cluster. That there was not leads back to the conclusion that southern approaches to biofuel policy making and the systems used to shape the biofuel policy arena are more complex than conventional wisdom suggests.

This diversity is also indicative of how each southern biofuel policy system adapts to the unique needs of its environment. Each southern state has developed biofuel policy portfolios according to its own culture, needs, and resources. It appears that the states that are the most reliant on agriculture—Arkansas, Mississippi, and Alabama—are those states with the fewest developed biofuel polices among southern states while those with less at stake agriculturally have developed more policies both in number and complexity, and with varying levels of governmental intervention. What has pushed this development? From a strictly rational policy-making perspective, those with the most at stake should have the more highly developed policies. If the states of Arkansas, Mississippi, and Alabama have the most to gain from biofuel policy as a result of well-developed agricultural capacity, why has the biofuel policy infrastructure not developed? What does this represent?

Looking beyond a simple rational perspective of the policy-making process and examining the policy outcomes and environment through various theoretical lenses, this pattern could be interpreted as the influence of traditional groups within the agriculture and business community that support status quo policies and eschew government interference in private agricultural business (Elazar 1984). It could also represent a lack of opportunity or "opening of a window" for the development of biofuel policy within the state political environment (Kingdon 2003), or it could simply mean that some states with fewer resources, such as those states that are heavily dependent on agriculture and highly rural, develop fewer policies while those with more develop more (Peters 2004; Sabatier 2007). Or there could be a more complex explanation that accounts for the development of a policy system through evolutionary processes representing adaptive and complex systems that have yet to be adequately addressed in theory.

However, comparisons made to other states in the union could lead to the conclusion that biofuel policy patterns and systems throughout the South mimic those same systems in other parts of the country, with the characteristics of each of these clusters found in other states in similar policy circumstances. Or conversely, the patterns developed in the South could be unique to its environment, agriculture, resources, and political traditions. Certainly both of these possibilities are worthy of further exploration, but attempts to do so without an understanding of how policy develops within a complex system of interest groups, policy makers, citizens, social and political cultures, and state level traditions will only develop unidimensional policy theories and not yield the requisite richness necessary to develop a real and valuable understanding of the policy process.

REFERENCES

Aldenderfer, M. S., and R. K. Blashfield. 1984. *Cluster Analysis*. Newbury Park, CA: Sage.

Alternative Fuels and Advanced Vehicles Data Center. 2008. State and Federal Incentives and Laws. *United States Department of Energy*. www.afdc.energy.gov (accessed on July 1, 2008).

American Farm Bureau Federation. 2009. Energy Tax Incentives. *American Farm Bureau Federation, 2008*. www.fb.org/issues/docs/renewabletaxincentives09.pdf (accessed on September 1, 2009).

Antares Group, Inc. 2007. U.S. Biofuel Production Potential. *Northeast Regional Biomass Program*. www.nrbp.org (accessed on May 1, 2008).

Badger, P. 2003. *Industry Survey Final Report: Developing State Policies Supportive of Bioenergy Development*. Norcross, GA: Southern States Energy Board.

Bailey, K. D. 1994. *Typologies and Taxonomies: An Introduction to Classification Techniques*. Thousand Oaks, CA: Sage.

Beam, D. R., and T. J. Conlan. 2002. Grants. In *The Tools of Government: A Guide to the New Governance*, edited by L. M. Salamon. New York: Oxford University Press.

Blair, R. 2002. Policy Tools Theory and Implementation Networks: Understanding State Enterprise Zone Partnerships. *Journal of Public Administration Research and Theory* 12, 2:161–90.

Bodman, S. 2008. Energy Information Administration 2008 Energy Conference. In *Remarks as Prepared for Delivery by Secretary Bodman*, edited by United States Department of Energy. Washington, DC.

Bordetsky, A., S. Casey-Lefkowitz, D. Lovaas, E. Martin-Perera, M. Nakagawa, B. Randall, and D. Woynillowicz. 2007. *Driving It Home: Choosing the Right Path for Fueling North America's Transportation Future*. New York: National Resources Defense Council.

Bureau of Economic Analysis. 2008. *Gross State Product by All Industries*. Washington, DC: United States Department of Commerce.

Bush, G. W. 2007. 2007 State of the Union Policy Initiatives. *Whitehouse 2007*. www.whitehouse.gov (accessed January 26, 2007).

Castles, F. G., and H. Obinger. 2008. Worlds, Families, Regimes: Country Clusters in European and OECD Are Public Policy. *West European Politics* 31, 1-2:321–44.

Cho, C. L., and D. S. Wright. 2001. Managing Carrots and Sticks: Changes in State Administrators' Perceptions of Cooperative and Coercive Federalism During the 1990s. *Publius—The Journal of Federalism* 31, 2:57–80.

Clemmer, S., B. Paulos, and A. Nogee. 2000. *Clean Power Surge: Ranking the States*. Cambridge, MA: Union of Concerned Scientists.

Common Purpose Institute. 2006a. Quick Facts on Biomass Energy Crops & Why We Need to Use Them in the South. 2006. *Common Purpose Institute*. www.treepower.org (accessed on January 29, 2006).

———. 2006b. Energy and the Environment. *Common Purpose Institute.* www.treepower .org (accessed September 28, 2006).

Cordes, J. J. 2002. Corrective Taxes, Charges, and Tradable Permits. In *The Tools of Government: A Guide to the New Governance*, edited byL. M. Salamon. New York: Oxford University Press.

Coyne, Marty. 2008. Energy Institute Applauds Call for Bold Federal Action. *Institute for 21st Century Energy, an Affiliate of the U.S. Chamber of Commerce.* www .energyxxi.org/xxi/newsroom/pr_080509 (accessed August 1, 2008).

Defenders of Wildlife. 2008. Getting Biofuels Right for Wildlife. *Defenders of Wildlife.* www.ewg.org/ (accessed May 1, 2008).

Derthick, M., ed. 2001. *Keeping the Compound Republic: Essays on American Federalism.* Washington, DC: Brookings Institution Press.

Elazar, D. J. 1984. *The American Mosaic: The Impact of Space, Time, and Culture on American Politics.* Boulder, CO: Westview Press.

———. 2001. Federalism: Collaboration and/or Its Evolution. *Governance—An International Journal of Policy and Administration* 14, 1:135–39.

Energy Efficiency and Renewable Energy Biomass Program. 2008. DOE and USDA Award $10 Million for Cellulosic Biofuel Research. *United States Department of Energy.* http://apps1.eere.energy.gov/news/news_detail.cfm/news_id=11903 (accessed September 19, 2008).

Energy Information Administration. 2007. Annual Energy Outlook 2007 with Projections to 2030 (early release). *United States Department of Energy.* www.eia.doe. gov/ (accessed August 1, 2008).

———. 2008. Annual Energy Outlook 2008 Overview. *United States Department of Energy.* www.eia.doe.gov/ (accessed August 1, 2008).

English, B. C., D. G. De La Torre Ugarte, K. Jensen, C. Hellwinkel, J. Manard, B. Wilson, R. Roberts, and M. Walsh. 2006. *25% Renewable Energy for the United States by 2025: Agricultural and Economic Impacts.* Knoxville, TN: The University of Tennessee.

Environmental Protection Agency. 2003. CO2 Emissions from Fossil Fuel Combustion Million Metric Tons CO2 (MMTCO2). *Environmental Protection Agency.* www.epa.gov/climatechange/emissions/downloads/CO2FFC_2003.xls (accessed February 1, 2007).

Environmental Working Group. 2007. *The Unintended Environmental Impacts of the Current Renewable Fuel Standard (RFS).* Washington, DC: Environmental Working Group.

Federal Highway Administration. 2008. Freight Management and Operations. *U.S. Department of Transportation.* http://ops.fhwa.dot.gov/freight/technology/index .htm (accessed May 1, 2008).

Feldman, R. J. 2002. Government Insurance. In *The Tools of Government: A Guide to the New Governance*, edited by L. M. Salamon. New York: Oxford University Press.

Greenberg, G., J. Miller, L. Morh, and B. Vladeck. 1977. Developing Public Policy Theory: Perspectives from Empirical Research. *American Political Science Review* 71, 4:1532–43.

Howard, C. 2002. Tax Expenditures. In *The Tools of Government: A Guide to the New Governance*, edited by L. M. Salamon. New York: Oxford University Press.

Ingram, H., A. L. Schneider, and P. DeLeon. 2007. Social Construction and Policy Design. In *Theories of the Policy Process*, edited by P. A. Sabatier. Cambridge, MA: Westview Press.

Institute for 21st Century Energy. 2008. Blueprint for Securing America's Energy Future. *The U.S. Chamber of Commerce.* http://www.energyxxi.org/reports/Blue_Print.pdf (accessed February 1, 2009).

Kelman, S. J. 2002. Contracting. In *The Tools of Government: A Guide to the New Governance*, edited by L. M. Salamon. New York: Oxford University Press.

Kettl, D. F. 1994. *Reinventing Government? Appraising the National Performance Review.* Washington, DC: The Brookings Institution.

Kettl, D. F., and J. J. DiIulio. 1995. *Inside the Reinvention Machine: Appraising Governmental Reform.* Washington, DC: The Brookings Institution.

Key, Jr., V. O. 1949. *Southern Politics.* New York: Vintage Books.

Kingdon, J. W. 2003. *Agendas, Alternatives and Public Policies.* New York: Longman.

Koonan, S. E. 2006. Getting Serious about Biofuels. *Science* 311, 5760:435.

Kraft, M. E., and S. R. Furlong. 2006. *Public Policy: Politics, Analysis, and Alternatives.* Washington, DC: CQ Press.

Lascoumes, P., and P. LeGales. 2007. Introduction: Understanding Public Policy through Its Instruments—From the Nature of Instruments to the Sociology of Public Policy Instrumentation. *Governance: an International Journal of Policy, Administration, and Institutions* 20, 1:1–21.

Leman, C. K. 2002. Direct Government. In *The Tools of Government: A Guide to the New Governance*, edited by L. M. Salamon. New York: Oxford University Press.

Linder, S. H., and B. G. Peters. 1981. Instruments of Government: Perceptions and Contexts. *Journal of Public Policy* 9, 1:35–58.

———. 1984. From Social Theory to Policy Design. *Journal of Public Policy* 4, 3:237–59.

Lowi, T. J. 1972. Four Systems of Policy, Politics, and Choice. *Public Administration Review* 32, 4:298–310.

Lubben, B. D., N. L. Bills, J. B. Johnson, and J. L. Novak. 2006. *The 2007 Farm Bill: U.S. Producer Preferences for Agricultural, Food and Public Policy.* Lincoln, NE: Farm Foundation.

MacDonald, T., G. Yowell, and M. McCormack. 2001. *U.S. Ethanol Industry: Production Capacity Outlook.* Sacramento, CA: California Energy Commission.

May, P. J. 2002. Social Regulation. In *The Tools of Government: A Guide to the New Governance*, edited by L. M. Salamon. New York: Oxford University Press.

Mazzocchi, M. 2008. SPSS Tutorial: AEB 37/AE 802, Marketing Research Methods, Week 7 *University of Brighton, UK.* www.personal.rdg.ac.uk/~aes02mm/ (accessed August 9, 2008).

McCoy, K. 2007. NAM Comprehensive Legislative Proposal: Energy Security for American Competitiveness. Washington, DC: National Association of Manufacturers. www.nam.org (accessed February 14, 2008).

Merriam-Webster. 2008. Merriam-Webster Online Dictionary. In *Merriam-Webster Online.* www.merriam-webster.com.

National Commission on the State and Local Public Service. 1993. *Hard Truths/Tough Choices: An Agenda for State and Local Reform.* Albany, NY: The Nelson A. Rockefeller Institute of Government.

Natural Resources Defense Council. 2007. *Getting Biofuels Right: Eight Steps for Reaping Real Environmental Benefits from Biofuels.* Washington, DC: Natural Resources Defense Council.

Obinger, H., and U. Wagschal. 2001. Families of Nations and Public Policy. *West European Politics* 24, 1:99–114.

Osborne, D., and T. Gaebler. 1992. *Reinventing Government: How the Entrepreneurial Spirit Is Transforming the Public Sector.* Reading, MA: Addison-Wesley.

Ostrom, V. 1989. *The Intellectual Crisis in American Public Administration.* Tuscaloosa, AL: The University of Alabama Press.

Peters, B. G. 2000. Policy Instruments and Public Management: Bridging the Gaps. *Journal of Public Administration Research and Theory* 10, 1:35–47.

———. 2004. *American Public Policy: Promise and Performance.* Washington, DC: CQ Press.

Peters, B. G., and S. H. Linder. 1998. The Study of Policy Instruments: Four Schools of Thought. In *Public Policy Instruments,* edited by B. G. Peters and F. van Nispen. Northhampton, MA: Edward Elgar Publishing, Inc.

Pickens, T. B. 2008. The Pickens Plan. 2008. *The Pickens Plan.* www.pickensplan.com/theplan/ (accessed September 20, 2008).

Sabatier, P. A. 2007. *Theories of the Policy Process.* Cambridge, MA: Westview Press.

Salamon, L. M. 1981. Rethinking Public Management: Third-Party Government and the Changing Forms of Government Action. *Public Policy* 29, 3:255–75.

———. 2002. The New Governance and the Tools of Public Action: An Introduction. In *The Tools of Government: A Guide to the New Governance,* edited by L. M. Salamon. New York: Oxford University Press.

Schneider, A. L., and H. Ingram. 1990. Behavioral Assumptions of Policy Tools. *Journal of Politics* 52, 2:510–29.

———. 1993. Social Construction of Target Populations: Implications for Politics and Policy. *The American Political Science Review* 87, 2:334–47.

———. 1997. *Policy Design for Democracy.* Lawrence, KS: University Press of Kansas.

Shafritz, J. M., and C. P. Borick. 2008. *Introducing Public Policy.* New York: Pearson Longman.

Sharkansky, I. 1970. *Regionalism in American Politics.* Indianapolis, IN: The Bobbs-Merrill Company, Inc.

Smith, K. B. 2002. Typologies, Taxonomies, and the Benefits of Policy Classification. *Policy Studies Journal* 30, 1:379–95.

Southern States Energy Board. 2005. *Outreach Support for Biomass Project Development in Florida: Value Added Metrics.* Norcross, GA: Southern States Energy Board.

Stanton, T. H. 2002. Loans and Loan Guarantees. In *The Tools of Government: A Guide to the New Governance,* edited by L. M. Salamon. New York: Oxford University Press.

Stanton, T. H., and R. C. Moe. 2002. Government Corporations and Government-Sponsored Enterprises. In *The Tools of Government: A Guide to the New Governance,* edited by L. M. Salamon. New York: Oxford University Press.

Steinberger, P. J. 1980. Typologies of Public Policy: Meaning Construction and Their Policy Process. *Social Science Quarterly* 61, 1:185–97.

Steuerle, C. E., and E. C. Twombly. 2002. Vouchers. In *The Tools of Government: A Guide to the New Governance,* edited by L. M. Salamon. New York: Oxford University Press.

Tolbert, V. R., and M. Downing. 1995. *Environmental Effects of Planting Biomass Crops at Larger Scales on Agricultural Lands.* Golden, CO: Second Biomass Conference of the Americas: Energy, Environment, Agriculture and Industry.

United States Department of Agriculture. 2008. *2007 Farm Bill, Title IX Energy.* U.S. Department of Agriculture [cited April 2008]. Available from www.usda.gov/documents/07title9.pdf.

United States Department of Energy. 2008. *Overview: Reliable, Affordable, and Environmentally Sound Energy for America's Future.* U.S. Department of Energy [cited April 2008]. Available from www.energy.gov/about/nationalenergypolicy.htm.

———. 2009. Secretary Chu Announces Nearly $800 Million from Recovery Act to Accelerate Biofuels Research and Commercialization. *United States Department of Energy.* www.energy.gov/news2009/7375.htm (accessed February 5, 2010).

United States Public Interest Group. 2007. The Carbon Boom: National and State Trends in Carbon Dioxide Emissions Since 1960. *United States Public Interest Research Groups.* www.uspirg.org (accessed January 5, 2007).

Wald, M. L. 2007. Is Ethanol for the Long Haul. *Scientific American,* 296, 1: 42–49.

Weber, M. 1958. *From Max Weber: Essays in Sociology.* Translated by H. H. Gerth and C. W. Mills. New York: Oxford University Press.

Weiss, J. A. 2002. Public Information. In *The Tools of Government: A Guide to the New Governance,* edited by L. M. Salamon. New York: Oxford University Press.

Westcott, P. C. 2007. U.S. Ethanol Expansion Driving Changes Throughout the Agricultural Sector. United States Department of Agriculture. www.ers.usda.gov/AmberWaves/September07/Features/Ethanol.htm (accessed September 5, 2008).

The White House. 2010. President Obama Announces Steps to Support Sustainable Energy Options, Department of Agriculture and Energy, Environmental Protection Agency to Lead Efforts. *White House.* www.whitehouse.gov (accessed February 4, 2010).

11

Distinctively South

Lessons for the Future of Environmental Management and Policy Implementation

John C. Morris and Gerald Andrews Emison

The purpose of this volume has been to examine in some detail the response of southern states to environmental policy imperatives. For many years the South has carried a reputation, often deserved, for being policy laggards, particularly in regards to national policy initiatives. Through the years a dizzying array of explanations has been offered, from V. O. Key's (1949) notions of states' rights, to Elazar's (1972) political culture, to Gray's (1969) policy innovation scores. Even in regards to environmental policy, Davis and Lester (1989) conclude that the South tends to lag far behind the rest of the nation when it comes to protecting the environment.

This collection of essays suggests, however, that these explanations are incomplete, in that they fail to account for the environmental innovation empirically evident in southern states. The South has undergone significant political change in the past thirty years (see Steed, Moreland, and Baker 1990), spurred by larger patterns of migration from "Rust Belt" states toward southern (and southwestern) states. With this growth in population comes a new set of environmental challenges. Disputes over water supply are generally associated with the arid West and Southwest, yet increasing demands for drinking water in expanding metropolitan regions of the South raises very similar issues. Likewise, air pollution, normally thought to be associated with cities of the industrial North or the automobile-choked West, is now common in many southern cities. In short, people change, lifestyles change, and politics change. At the same time, we make no claim of revolutionary change; traditions and history are still important elements of southern culture (Carmines and Stanley 1990; Key 1949), and in many ways, these elements are still very relevant to understanding environmental policy in the South. Yet there is empirical evidence of change that cannot be ignored—change that cannot be explained through existing frameworks.

Moreover, this same innovation is marked by conflict. While the South is often thought of as a relatively homogeneous society, there are different values and interests that lead to conflict. Social conflict is no stranger to the South; in fact, the history of the South is marred by conflict in many forms. It should come as no surprise, then, that conflict is also present in environmental policy. By understanding the terms of this conflict, we can begin to better understand the unique mix of choices states make in this policy realm.

THE NATURE OF CONFLICT

The conflict in each of these cases manifests itself in different ways, and for different reasons. In some cases, the dispute is grounded in a conflict of interests between political entities and/or interest groups. In other cases, conflict is created by resource shortages, perceived or real; in still other instances the conflict is driven by more fundamental differences between the public and private sectors. In the field of environmental management, an overarching level of conflict may be found in the discord between traditional command-and-control environmental management and postmodern environmental management.

Conflicts of Interest

Conflicts of interest are a central feature of the American political system and are as important in the South as in any other region of the nation. People have different interests; the expression of these interests in the policy realm thus creates political conflict. This conflict shapes, in turn, the policy choices available to states, and defines the conditions under which states can (or will) react to national environmental mandates. The differences highlighted in the Clean Water State Revolving Fund program between Georgia, Alabama, and Mississippi provide a case in point; each state ultimately adopted very different enabling legislation, and thus program structures, because the mix of interests present in each state was unique. Likewise, North Carolina and Florida developed different approaches to the problem of brownfields for the same reasons.

Resources Shortages

Nearly every state government in the nation has experienced revenue shortfalls in recent years. State legislatures are forced to make hard decisions about how to appropriate the available funds, and when resources are scarce, environmental imperatives tend to creep lower on the list of budget priorities. At the same time, a lack of resources does not lessen the pressure by citizens and interest groups to meet environmental goals, nor does it lessen the pressure on state environmental agencies to innovate. Even in times of relative fiscal comfort, state environmental agencies in the South may lack the political influence to ensure their activities and initiatives are fully funded. The Mississippi Department of Environmental Quality illustrates this issue: the agency is under increasing pressure to adopt new management processes and implement innovative programs, yet their efforts are increasingly hampered by a shortage of fiscal and political resources.

These shortages can also come in the form of physical resources in the environment; the dispute between Georgia, Alabama, and Florida over water supply in the Chattahoochee River Basin is a case in point. There is currently plenty of water in the river system to meet the needs of the communities in the watershed, yet battles persist over perceived future shortages, with each state seeking to protect its ability to grow its population in the region in the future. This conflict is also exacerbated by the potential environmental consequences of water diversion or water shortage in the watershed.

Finally, a third type of resource shortage is represented by shortages in administrative capacity. An index of administrative capacity developed by Bowman and Kearney (1988) generally ranks southern states among the lowest in the nation, yet these states are under the same federal legislative mandates as the rest of the country. Alabama and Mississippi each lacked the administrative and technical expertise to design, implement, and operate a complex environmental loan program as required by federal law, and each state adopted a different solution: Alabama sought private sector expertise, while Mississippi implemented a "bare bones" program. Both states were ultimately able to implement the program, but with very different results. Each state adopted a solution driven largely by the resources available to them.

Public-Private Conflict

The clash of values between the public and private sectors is an ongoing source of conflict in all states. A significant body of literature exists that examines the nature of this conflict (see, for example, Donahue 1989; Heilman and Johnson 1992; Kettl 1993; Savas 2000), yet conclusions about the outcome of this conflict vary widely. Donahue (1989) suggests that conflict is an inevitable outcome of public-private relationships, and that it is unlikely that such conflict can be resolved. On the other hand, Heilman and Johnson (1992) conclude that carefully planned, participative public-private partnerships can reduce conflict to manageable levels. Their mechanism for reaching this conclusion, however, is interesting:

> However, in the public-private partnership movement, privatization, at least successful privatization, is a paradox. Rather than reducing government, which it may do in some ways, more fundamentally it brings the private sector into the policy and management structures of the public sector. The private sector becomes more like the public sector than the reverse. (Heilman and Johnson 1992, 190)

While Heilman and Johnson reach this finding at the conclusion of a study of the privatization of wastewater treatment plants, it is interesting to note that conflict between the public and private sectors over public values such as transparency and oversight is very much a part of the brownfields programs in North Carolina and Florida. The lesson here may be that true public-private partnership is very difficult to achieve, yet without it, conflict is inevitable.

Command-and-Control vs. Postmodern Environmental Management

Each of the cases in this volume illustrates the tremendous complexity present in environmental management structures. At the heart of the issue is the traditional reliance on Weberian, command-and-control, authority-based organizational structures. Ubiquitous in governments of all descriptions, these structures are excellent mechanisms to promote certain larger societal and political values—stability, accountability, professionalism, transparency, and political neutrality. Basic administrative theory also tells us that increasing levels of complexity present in an administrative task will manifest itself as increasing levels of complexity in the administrative structure assigned to the task. Furthermore, organizational theory links the level of uncertainty in an interorganizational environment to the level of complexity in an organization.

Environmental problems are, by their very nature, complex and uncertain issues, in which causal relationships may be poorly understood, physical processes mysterious, and effects highly interrelated. Moreover, environmental problems are rarely confined to a single geographic area—effluent dumped in a river may pass through many political jurisdictions on its way to the sea; polluted air may likewise affect many population centers before dissipating. It should be no surprise, then, that complexity should also be found in state environmental agencies. Indeed, task specialization is a central feature of such agencies: there are separate offices for air quality programs, drinking water programs, wastewater programs, toxic and hazardous waste programs, and so on. This same complexity extends beyond the notional "borders" of environmental agencies; successful implementation of many environmental programs requires close cooperation between different agencies of state government, the national government, local governments, and even between state governments, thus invoking both horizontal and vertical federalism. When one stretches the membership to private and nonprofit organizations, the mix of incentives, goals, and values grows exponentially more complex.

If traditional bureaucratic management structures are employed to address these problems, several conflicts immediately become apparent. First, traditional bureaucratic structures rely on clear, nonoverlapping lines of authority, in which every player in the system reports to a single authority figure. Yet when multi-actor implementation structures are created to address environmental problems, lines of authority are often overlapping, unclear, or altogether missing. A classic implementation study by Pressman and Wildavsky (1973) also illustrates the tremendous capacity for altered goals, orders, and directives as existing authority mechanisms are stretched over time and distance. Second, when these overlapping authorities cross organizational boundaries, there must be some external authoritative figure to adjudicate the inevitable disputes. Bureaucratic squabbling, credit-claiming, or responsibility-shirking behavior are commonplace, and when present among otherwise separate and independent agencies, the ability of any single authoritative figure to compel compliance is limited. The failures of American intelligence agencies in the late twentieth and early twenty-first centuries are testament to this problem. Third, while the Constitution specifies the separation of powers in a vertical dimension, it is largely mute on the question of authoritative relationships between states and/or between local governments. The Constitution does specify a dispute resolution mechanism, but it is often a mechanism of last resort.

If southern states do lag behind the rest of the nation in terms of administrative capacity (Bowman and Kearney 1988), and the dominant political culture in the South is traditionalistic (Elazar 1972), then it should be unremarkable that traditional bureaucratic management structures are present in these environmental agencies. This is not to say, however, that attempts to overcome these limitations are absent. To the contrary, the cases in this volume illustrate clearly that officials in these agencies are continuously searching for innovative solutions to environmental problems and, in some cases, succeeding. These same efforts, though, are often frustrated by a social, political, and economic environment that values the status quo and that continues to operate largely under a traditional command-and-control structure. The basis of the conflict, then, is between the entrenched command-and-control bureaucratic structures present in states and the emergent, dynamic, and uncertain nature of the policy problems addressed by these structures.

CONSIDERING THE COMPLEXITY OF
SOUTHERN ENVIRONMENTAL MANAGEMENT

One theme that characterizes the environmental management discussed in this volume is the complexity of such efforts. It might be helpful to explore this complexity in more detail. First, is it reasonable to view environmental management systems as "complex systems"? The qualified answer to this question is "yes." The cases in this volume all speak to the dynamic, complex, emergent nature of environmental problems, and of the systems created to address these problems. While one may view much of this effort as "muddling through" (Lindblom 1959) on the part of southern states, such a conclusion is at best incomplete and at worst disingenuous. Real innovation is taking place, but it is happening in a context that strongly resists innovation and change. We see evidence that environmental managers are operating within these constraints to find solutions to difficult problems, whether the issue is coastal zone management, brownfields reclamation, wastewater infrastructure financing, or addressing environmental justice.

Second, can the relationships between the components of the environmental management system be characterized simply, and can their interactions be predicted with precision? Our cases suggest that not only does the environmental management system defy simplicity, it also defies precision. As new environmental issues have emerged and been incorporated into existing structures, those structures have become more complex, just as organization theory would suggest. If one could indeed "simply characterize" southern environmental management systems, the simple characteristic would be "complex," thus begging the same question.

The larger questions of precision and predictability are not unique to environmental policy, nor are they unique to southern states. Indeed, these are issues that plague policy makers of all stripes everywhere. Nearly all policy solutions are causal ("if we do 'A,' then 'B' will be the result"), yet unintended policy consequences are ubiquitous. True policy precision and predictability require omniscience, a trait in short supply in the human race, particularly in the social and policy sciences. While causality is more often associated with the natural sciences, our experiences with the environmental sciences lead us to conclude that there is much we do not yet understand about these processes. In short, we have a scenario in which we may lack clear scientific guidance; we lack predictive ability; we are heavily invested in an incompatible organizational and management structure; and we must account for a wide panorama of larger social, political, and economic interests. It should thus come as no surprise that precision and predictability are lacking. All of the players involved work hard on the problem, but outcomes are just as often largely determined by luck as they are knowledge and foresight.

Finally, does the environmental management system display the property of emergence? There is ample evidence to suggest it does. Some of the changes are driven by larger changes in metapolicy (Johnson and Heilman 1987), some are driven by advances in our understanding of the science behind environmental policy, and some are driven by changes in the context (political, social, and economic) that open new paths and choices. At the same time, these changes, and their consequences, are largely impossible to predict. The net result of this state of affairs is that not only is environmental management *emergent*, it is also largely *reactive*, much as an emergency room is reactive—it seeks to remediate the effects of past events, not prevent future harm.

Thus environmental managers are forced to address problems already present. Since change comes slowly in traditional bureaucratic structures, these existing problems have additional time to worsen before we can bring our attention and resources to bear on the problem. When resource shortages limit the ability of environmental managers to address these problems, the level of conflict rises even further.

Hope for the Future?

One might possibly conclude from this discussion that not only are southern states doomed to failure, but that our ability as a nation to address environmental imperatives is seriously in question. Regardless of the challenges before us, our reading of the evidence from these cases suggests quite the opposite. In spite of the significant barriers in place, environmental managers in the South show an almost remarkable resiliency and ability to innovate within the constraints present. The path of least resistance, clearly, would be for environmental managers to expend the least effort necessary to meet national policy requirements and not to seek change. In some ways, the unique contextual features of the South raise the "level of difficulty" for southern environmental managers, forcing them to be more creative in the ways they seek to meet these requirements. This is not to say that environmental managers in other regions of the country do not face significant barriers; in many ways, the barriers present in the South are much like those in the rest of the nation. Still, those barriers are different, and they differ in important ways. That we see evidence of innovation and change is testament to these collective efforts.

DIRECTIONS FOR FUTURE RESEARCH

This volume has examined the issues of exceptionalism and innovation in southern states. Our research has identified several key variables important to the broader questions of innovation in environmental management.

Chief among these is the context in which the management activity takes place. Related to context are the variables of resources, public-private value conflicts, existing organizational complexity, problem complexity, and the conflict of political interests. It remains the subject of future research to better specify these variables, operationalize them, and incorporate them into a broader theory of environmental management.

While existing state comparative literature may provide some guidance for this effort, we believe an inductive approach to theory building is equally useful in this endeavor. An enduring problem of policy research is the "cases/variables" problem (Goggin 1986), and that problem is present here as well. Thus, it may be that the topic lends itself more readily to qualitative case study methodologies rather than quantitative methods more suited to "large-N" studies. In any event, efforts to better specify the important variables as components of the theory-building process will reveal the most fruitful methodological approaches.

Likewise, there is a need for better specification of a theory of the management of complex administrative systems. As noted throughout this discussion, previous work has attempted to describe, explain, and even predict administrative behavior in complex environments, but most of these efforts are constrained by their inherent assumptions of known, testable causal relationships. The present approach begins from a different premise, one in which assumptions of known causality are inconsistent. While we are skeptical of our ability to develop high-level (predictive) theory to model this complexity, we are confident that mid-level (explanatory) theory is both eminently possible and fruitful.

Other areas ripe for future work include an effort to examine these policy areas across all states of the South simultaneously. While we know something about environmental justice issues in Florida, for example, or infrastructure financing in Georgia, Alabama, and Mississippi, the degree to which these issues are replicated in other southern states is a question requiring empirical study. We have every reason to believe they will play out similarly in other locations, but only through careful examination can this proposition be confirmed. Likewise, we believe it is highly likely that similar issues are present in other regions of the nation. States in the West battle regularly over water rights; states in the Chesapeake Bay watershed are attempting to address complex pollution issues stretching across thousands of square miles; states in the Mississippi River Basin share water pollution and water use problems; and states in the industrial Midwest battle over issues of air pollution. All states operate under the same federal policy mandates, yet each state (and each region) is fully interconnected with its own context.

There is much we know about environmental management, and much we do not know. Our efforts here are focused on the empirical questions (e.g., what are the circumstances?), and explicitly not on the normative questions (e.g., what should be the circumstances?, or is complex adaptive management better than traditional bureaucratic management?). These questions are not unimportant, but they are more reasonably within the purview of the policy system itself. To challenge the foundation of that same policy system, that causal relationships are known and predictable, is likely to be a difficult task, but may ultimately lead to more effective environmental management.

REFERENCES

Bowman, A. O'M., and R. C. Kearney. 1988. Dimensions of State Government Capability. *Western Political Quarterly*, 41:341–62.

Carmines, E. G., and H. W. Stanley. 1990. Ideological Realignment in the South: Where Have all the Conservatives Gone? In *The Disappearing South? Studies in Regional Change and Continuity*, ed. R. P. Steed, L. W. Moreland, and T. A. Baker, 5–20. Tuscaloosa, AL: University of Alabama Press.

Davis, C. E., and J. P. Lester. 1989. Federalism and Environmental Policy. In *Environmental Politics and Policy: Theories and Evidence*, edited by J. P. Lester, 57–84. Durham, NC: Duke University Press.

Donahue, J. P. 1989. *The Privatization Decision: Public Ends, Private Means*. New York: Basic Books.

Elazar, D. J. 1972. *American Federalism: A View from the States*, 2nd ed. New York: Thomas Crowell.

Goggin, M. 1986. The "Too Few Cases/Too Many Variables" Problem in Implementation Research. *Western Political Quarterly* 38:328–47.

Gray, V. 1969. Innovation in the States: A Diffusion Study. *American Political Science Review* 67:1174–85.

Heilman, J. G., and G. W. Johnson. 1992. *The Politics and Economics of Privatization: The Case of Wastewater Treatment*. Tuscaloosa, AL: University of Alabama Press.

Johnson, G. W., and J. G. Heilman. 1987. Metapolicy Transition and Policy Implementation: New Federalism and Privatization. *Public Administration Review* 47:468–78.

Kettl, D. F. 1993. *Sharing Power: Public Governance and Private Markets*. Washington, DC: Brookings.

Key, V. O. 1949. *Southern Politics in State and Nation*. New York: Alfred P. Knopf.

Lindblom, C. E. 1959. The Science of Muddling Through. *Public Administration Review* 19:79–88.

Pressman, J., and A. Wildavsky. 1973. *Implementation: How Great Expectations in Washington Are Dashed in Oakland; Or, Why It's Amazing That Federal Programs Work at All, This Being a Saga of the Economic Development Administration as Told by Two Sympathetic Observers Who Seek to Build Morals on a Foundation of Ruined Hopes*. Berkeley, CA: University of California Press.

Savas, E. S. 2000. *Privatization and Psublic-Private Partnerships*. New York: Chatham House.

Steed, R. P., L. W. Moreland, and T. A. Baker, eds. 1990. *The Disappearing South? Studies in Regional Change and Continuity*. Tuscaloosa, AL: University of Alabama Press.

About the Contributors

David A. Breaux is professor of political science and public administration and associate dean of the College of Arts and Sciences at Mississippi State University. He earned his BA from Nicholls State University (1980), his MA from the University of New Orleans (1984), and his PhD from the University of Kentucky (1989). His teaching and research expertise is in the area of state politics and policy. He has published articles in various peer-reviewed journals, including *Legislative Studies Quarterly*, *American Politics Quarterly*, *American Review of Politics*, and *Public Administration Review*, and has also published numerous book chapters.

Gerald Andrews Emison is an associate professor of political science and public administration at Mississippi State University. His research interests concern effectiveness of public environmental institutions, professionalism in city planning and engineering, and environmental consequences of growth management. He is the author of *Practical Program Evaluations: Getting from Ideas to Outcomes* (2006) as well as over twenty journal articles/book chapters. He is retired from the senior executive service of the U.S. Environmental Protection Agency. He is a registered professional engineer, a diplomat of the American Academy of Environmental Engineers, and a member of the American Institute of Certified Planners.

Deborah R. Gallagher is assistant professor of the practice of environmental policy in Duke University's Nicholas School of the Environment and executive director of the Duke Environmental Leadership program. She received her PhD from the University of North Carolina at Chapel Hill in 2002. Her research focuses on public policies governing business and the environment, environmentally sustainable strategic management, and the role of business in global environmental governance. Her work has been published in *Journal of Environmental Planning and Management* and *Local Environment*, and in collections such as *Organizations and the Sustainability Mosaic* and *New Horizons in Research in Sustainable Organizations*.

Celeste Murphy-Greene has published several journal articles and book chapters appearing in such publications as *Public Administration Review, Review of Policy Research, Public Administration Quarterly, International Journal of Public Administration, Journal of Health and Human Services Administration,* and the *Journal of Public Management and Social Policy*. She is currently the program coordinator and an adjunct professor for the University of Virginia's Graduate Certificate in Public Administration and also teaches as an adjunct Professor at the Old Dominion University. She earned her PhD in public administration from Florida Atlantic University.

Erin Holmes is the director of institutional research and assessment at Black Hills State University. She received her PhD in public policy and public administration from Mississippi State University in 2009. Her research interests include policy instruments, federalism, political culture, and higher education policy.

Madeleine W. McNamara is a senior administrator for the Coast Guard and an adjunct professor at the University of New Orleans. She is a graduate of the U.S. Coast Guard Academy and received her PhD from Old Dominion University in 2008. Her research interests include collaboration, public policy, environmental management, and networks. She has published in *Public Works Management and Policy* and *Virginia Social Science Journal*.

John C. Morris is a professor of public policy and public administration in the Department of Urban Studies and Public Administration at Old Dominion University. He received his PhD from Auburn University in 1994. His research interests include public-private partnerships, federalism, public policy, and collaboration. He has published over fifty scholarly articles and book chapters, including work in journals such as *Public Administration Review, Journal of Politics, Environmental Politics, Policy Studies Journal,* and *State and Local Government Review,* among others. He is also coeditor of *Building the Local Economy: Cases in Economic Development* (2008).

James Newman is an assistant professor of public administration in the political science department at Idaho State University. He received his PhD from Mississippi State University in 2006. His research interests include intergovernmental relations, water allocation policy, and evaluation for nonprofit organizations. He has recently completed a book about the ACT-ACF compact negotiations.

Rick Travis is an associate professor of political science at Mississippi State University. His research focuses on public policy and foreign policy making. He received his PhD from the University of Georgia in 1993. His work has been published in *International Studies Quarterly, Policy Studies Journal, International Interactions, Social Science Quarterly,* and elsewhere.

Breinigsville, PA USA
13 April 2011
259723BV00003B/21/P